Plant Genomics and Climate Change

David Edwards • Jacqueline Batley
Editors

Plant Genomics and Climate Change

Editors
David Edwards
School of Plant Biology and Institute
of Agriculture
University of Western Australia
Crawley, WA, Australia

Jacqueline Batley
School of Plant Biology and Institute
of Agriculture
University of Western Australia
Crawley, WA, Australia

ISBN 978-1-4939-3534-5 ISBN 978-1-4939-3536-9 (eBook)
DOI 10.1007/978-1-4939-3536-9

Library of Congress Control Number: 2016933330

Springer New York Heidelberg Dordrecht London

Printed on acid-free paper

Springer Science+Business Media LLC New York is part of Springer Science+Business Media (www.springer.com)

Preface

The sustainability of agriculture is being challenged by climate change and rising food demand from a larger and wealthier human population. Humanity faces a global food deficit unless the efficiency and resilience of crop production is improved.

Within the coming decades challenges to international food production will occur like no other time in human history, and a substantial increase in the production of food is essential if we are to continue to feed the growing human population. There is an urgent need to increase crop yield, quality and stability of production, enhancing the resilience of crops to climate variability and increasing the productivity of minor crops to diversify food production.

Improvements in agricultural practice and the increased use of fertilisers and pesticides have increased food production over the last few decades; however it is now considered that further such improvements are limited. The science of genomics offers the greatest potential for crop improvement.

This book explores the impact of climate change on agriculture and our future ability to produce the crops which are the foundation of the human diet. Further chapters address the specific climate change issues and explore the potential for genomics-assisted breeding of improved crops with greater yield and tolerance to the stresses associated with predicted climate change scenarios.

Through the application of genomics technology, it is possible to accelerate the breeding of major crops, bring current orphan crops into accelerated agricultural breeding programs and convert diverse non-crop species into future crops adapted to the changing climate. Through this process we can help secure the food supply for the coming generations.

David Edwards
Jacqueline Batley

Contents

Authors Bios

Prof David Edwards gained an Honours degree in Agricultural Science from the University of Nottingham and a PhD from the Department of Plant Sciences, University of Cambridge. He has held positions within academia (University of Adelaide and University of Queensland, Australia; University of Cambridge, UK; and McGill University, Canada), government (Long Ashton Research Centre, UK, Department of Primary Industries, Victoria, Australia) and industry (ICI seeds, UK). David was appointed as a Centenary Professor at The University of Western Australia in 2015. His research interests include the structure and expression of plant genomes, the discovery and application of genome variation and applied bioinformatics, with a focus on crop plants and accelerating crop improvement in the face of climate change.

Prof Jacqueline Batley is an ARC Future Fellow at the University of Western Australia. She was awarded her PhD from the University of Bristol in 2001 and moved to Australia in 2002. Jacqueline has expertise in the fields of plant and animal molecular biology, genetics and genomics, gained from working in both industry and academia. Her areas of interest include genetic and genomic analysis for applications including genetic diversity, linkage disequilibrium and comparative genomic studies, working across environmental and agricultural areas. Her current research projects include the molecular characterisation of agronomic traits, with a focus on disease resistance in Brassicas, with studies in both the fungal pathogen and the host plant.

Contributors

Isabel A. Abreu, Ph.D. Instituto de Tecnologia Química e Biológica António Xavier, Universidade Nova de Lisboa, Oeiras, Portugal

Carla António, Ph.D. Instituto de Tecnologia Química e Biológica António Xavier, Universidade Nova de Lisboa, Oeiras, Portugal

Bruna C. Arenque, M.Sc. Botany Department, University of São Paulo, Sao Paulo, Brazil

Pedro M. Barros, Ph.D. Instituto de Tecnologia Química e Biológica António Xavier, Universidade Nova de Lisboa, Oeiras, Portugal

Philipp Emanuel Bayer, B.Sc., Ph.D. School of Agriculture and Plant Sciences, University of Queensland, Brisbane, QLD, Australia

Perrin H. Beatty, Ph.D. Department of Biological Sciences, University of Alberta, Edmonton, AB, Canada

Marcos S. Buckeridge, Ph.D. Botany Department, University of São Paulo, Sao Paulo, Brazil

David Edwards, Ph.D. School of Plant Biology and Institute of Agriculture, University of Western Australia, Crawley, WA, Australia

Bruce D.L. Fitt, M.A., Ph.D., D.I.C., F.R.S.B. School of Life & Medical Sciences, University of Hertfordshire, Hatfield, Hertfordshire, UK

Timothy Fitzgerald, Ph.D. School of Plant Biology, University of Western Australia, Crawley, WA, Australia

Robert C. Godfree, Ph.D. National Research Collections Australia, Commonwealth Scientific and Industrial Research Organisation, Canberra, ACT, Australia

Allen Good, B.Sc., M.Sc., Ph.D. Department of Biological Sciences, University of Alberta, Edmonton, AB, Canada

Mei Han, Ph.D. Department of Biological Sciences, University of Alberta, Edmonton, AB, Canada

David J. Hughes, M.Sc., D.Phil. Computational & Systems Biology, Rothamsted Research, Hatfield, Hertfordshire, UK

Tiago F. Lourenço, Ph.D. Instituto de Tecnologia Química e Biológica António Xavier, Universidade Nova de Lisboa, Oeiras, Portugal

Paula Andrea Martinez, B.Sc. School of Agriculture and Plant Sciences, University of Queensland, Brisbane, QLD, Australia

Jennifer MingSuet Ng Department of Biological Sciences, University of Alberta, Edmonton, AB, Canada

M. Margarida Oliveira, Ph.D. Instituto de Tecnologia Química e Biológica António Xavier, Universidade Nova de Lisboa, Oeiras, Portugal

João S. Pereira, Ph.D. Instituto Superior de Agronomia, Universidade de Lisboa, Lisbon, Portugal

Jennifer C. Pierson, Ph.D. Fenner School of Environment and Society, The Australian National University, Acton, ACT, Australia

Nelson J.M. Saibo, Ph.D. Instituto de Tecnologia Química e Biológica António Xavier, Universidade Nova de Lisboa, Oeiras, Portugal

Ana Paula Santos, Ph.D. Instituto de Tecnologia Química e Biológica António Xavier, Universidade Nova de Lisboa, Oeiras, Portugal

Amanda P. De Souza, M.D. Botany Department, University of São Paulo, Sao Paulo, Brazil

Henrik U. Stotz, Ph.D. School of Life & Medical Sciences, University of Hertfordshire, Hatfield, Hertfordshire, UK

Eveline Q.P. Tavares, M.Sc. Botany Department, University of São Paulo, Sao Paulo, Brazil

Lyndsey M. Vivian, Ph.D. National Research Collections Australia, CSIRO, Canberra, ACT, Australia

The Impact of Climate Change on Agricultural Crops

Timothy Fitzgerald

Introduction

Human civilization is dependent on agricultural plant production. The vast majority of human calories are derived from agricultural crops either directly, or indirectly as animal feed (Cassidy et al. 2013). A relatively small number of crop species are cultivated on a large scale globally (Prescott-Allen and Prescott-Allen 1990) with the three cereals rice, wheat, and maize accounting for over 60 % of human energy intake alone (Cassman et al. 2003). Our major crops are grown across regions with diverse climates. However, to achieve this, decades of effort have been invested in the development of locally adapted varieties and in optimizing agricultural practices (e.g. timing of sowing and harvest; irrigation), and it is nevertheless true that for a given crop there exists a relatively small climatic window that is optimal for production (Lobell and Gourdji 2012).

Over the last few decades, increases in mean global surface temperature and the atmospheric concentration of greenhouse gases including carbon dioxide (CO_2) and ozone (O_3) have been observed. The Intergovernmental Panel on Climate Change (IPCC) reports that it is almost certain that these trends will continue throughout the twenty-first century (Stocker et al. 2013). There is substantial evidence that this will affect crop productivity (Lobell and Gourdji 2012), and with accelerating gains in crop yields essential to meet demands of an expanding population, food security is becoming one of the most important issues facing humanity.

Here I provide a summary of historic observations of climate change, and predictions for continued changes during the twenty-first century. I then outline current understanding of the effects of climate change (and the associated accumulation of CO_2 and O_3) on plant growth and development, highlighting projected impacts on

T. Fitzgerald, PhD (✉)
School of Plant Biology, University of Western Australia,
35 Stirling Highway, Crawley, WA 6009, Australia
e-mail: timothy.l.fitzgerald@gmail.com

D. Edwards, J. Batley (eds.), *Plant Genomics and Climate Change*,
DOI 10.1007/978-1-4939-3536-9_1

crop productivity. The effect of these factors on plant pathogens and insects (both pests and pollinators) and their interaction with crops, is also discussed. An overview of the potential for adaptation of crop production to a changing climate is provided, with both 'on-farm' and larger-scale options described. Finally, the contribution of crop production itself to climate change is summarised, and proposed strategies to mitigate this are outlined.

Observed and Predicted Climate Change as a Result of Human Activity

As presented in the latest Intergovernmental Panel on Climate Change (IPCC) report (Stocker et al. 2013), it has been demonstrated through multiple lines of evidence that the global climate is changing, and that human activities are a major driving force for this change. Climate change occurring as a result of human activity can be specifically referred to as 'anthropogenic climate change'. The terms 'climate change' and 'anthropogenic climate change' are frequently used interchangeably; however trends and fluctuations in climate independent of human activity are also well-known to occur. In addition to seasonal variations, several natural phenomena contribute to substantial short-term fluctuations in the climate (occurring over periods of years to decades), including the El Niño-Southern Oscillation (ENSO), the Pacific Decadal Oscillation (PDO), the North Atlantic Oscillation (NAO) (Rosenzweig and Neofotis 2013) and short term solar cycles (Keckhut et al. 2005). Furthermore, longer term climate fluctuations (occurring over centuries to millennia) are driven by factors that include variations in Earth's orbit (Spiegel et al. 2010) and long-term solar cycles (Wilson 2003). The presence of non-anthropogenic phenomena influencing climate change (and their potential interactions with anthropogenic factors) complicates predictions of future climate scenarios; however considerable international effort has been expended to develop models to account for these factors as accurately as possible.

The average surface temperature of Earth has increased by approximately 0.8 °C since the early twentieth century, and the majority of this warming has occurred since 1980 (Carnesale and Chameides 2011). Importantly, changes in surface temperature represent only a small proportion of global warming; 90 % of the Earth's warming since 1971 has occurred in the oceans (Stocker et al. 2013). The scientific consensus, incorporating analysis of both anthropogenic and non-anthropogenic factors influencing climate, is that the majority of this temperature increase is the result of greenhouse gas emissions from human activity (Stocker et al. 2013). Extensive international effort has been invested in the development of models to predict mean temperature increases during the twenty-first century, based on a range of emission scenarios. The most recent IPCC report states that it is 'likely' (>66 % probability) under most future greenhouse gas emission scenarios that the average global surface temperature will increase by more than 1.5 °C by 2100. In a scenario where emissions continue to increase, following historic trends, an average temperature increase of 3.7 °C is predicted (Stocker et al. 2013). Climate modelling and historical data suggest that surface temperature increase as a result of greenhouse gas emissions is more dramatic on land than at sea (Sutton et al. 2007), thus the increase in temperature to which humans and crops are exposed is likely to be greater than these projected averages.

In addition to the observed increase in mean global surface temperature, the early part of the twenty-first century has seen a very high frequency of extreme weather events (Coumou and Rahmstorf 2012) and several recent studies have suggested that these phenomena may be related (Trenberth 2012). The 2013 IPCC report has concluded that, in addition to a rise in mean global surface temperature, it is very likely (>90 % certainty) that by the end of the twenty-first century there will be an increase in the length and intensity of heatwaves, and the frequency and/or duration of heavy rainfall events.

Effects of Climate Change on Crop Productivity

Elevated Mean Surface Temperature

The C3, C4 and CAM photosynthetic pathways are differentially affected by temperature, and thus the effect of elevated temperature on the performance of a given crop is expected to be influenced by the type of photosynthesis it employs. Generally, the optimum temperature for photosynthesis is highest in C4 plants and lowest in CAM plants (Yamori et al. 2014). Broadly however, increased temperature exerts two main direct effects on plant growth and development. Firstly, higher temperatures accelerate development (Yin et al. 1995) resulting in a shorter duration of the cropping cycle in annual crops, which generally translates to lower yield (Wheeler et al. 2000; Porter and Semenov 2005). Secondly, elevated temperature affects photosynthetic rate; this may lead to a loss or gain in efficiency depending on whether the temperature moves closer to or further away from the optimum (Lobell and Gourdji 2012). In addition, changes in temperature exert an indirect effect on growth and development via changes in vapour pressure deficit (VPD) between air and leaf. For a set relative humidity, increases in temperature lead to an increase in vapour pressure deficit (Lobell and Gourdji 2012) resulting in reduced water use efficiency. Plants attempt to offset this by reducing stomatal aperture (Polley 2002), however this in turn leads to decreased rates of photosynthesis and an increase in crop canopy temperature (Leinonen and Jones 2004).

Increases in mean temperature may also have important effects on growing seasons throughout the world. In regions including Russia, Northern China, and Canada, there is expected to be a substantial increase in the frost-free period suitable for crop cultivation during the twenty-first century (Ramankutty et al. 2002). In contrast, in many temperate (including North America and Northern Asia) and tropical regions (including parts of South America, Australia and Asia) crops are more likely to be exposed to heat stress during growing seasons, which can lead to substantial decreases in productivity (Teixeira et al. 2013).

Previous reports have indicated that increases in temperature of up to 2 °C could exert an overall positive effect on global production of the staple cereal crops rice, wheat and maize, with additional temperature increases leading to productivity decreases (Parry 2007). However, a recent large-scale meta-analysis suggests that, with the exception of wheat grown in tropical regions (for which yield increases are predicted for increases in temperature of up to 1 °C) any increase in mean global surface temperature is likely to result in decreased productivity for these crops (Challinor et al. 2014).

Increased Frequency of Extreme Weather Events

An increase in the frequency of heatwaves and extreme rainfall events is predicted, with high confidence, for the twenty-first century (Stocker et al. 2013). Previous IPCC projections have indicated a likely increase in drought and cyclone activity, both of which could have major detrimental effects on global crop productivity (Schmidhuber and Tubiello 2007). However the most recent report incorporating the latest data and models, states a high degree of uncertainty with regard to these phenomena (Stocker et al. 2013).

Both heatwaves and extreme precipitation jeopardise crop yield. Extreme temperatures can damage plant cells, exerting a negative effect on growth and development. The reproductive phase of plant development is highly sensitive to temperature stress (Barnabás et al. 2008); extreme heat during this period can lead to sterility and dramatic decreases in yield. A recent study (Deryng et al. 2014) suggests a strong negative influence of extreme heat events on the productivity of maize, wheat, and soybean by the middle of the twenty-first century, under all likely emission scenarios. Extreme precipitation events can also cause major crop losses, via physical damage from flooding and excess soil moisture. The latter can cause anoxic conditions, leading to above and below ground damage (Kozdrój and van Elsas 2000), and increases the risk of certain plant diseases (Ashraf 1999). Furthermore, excessive rainfall can result in delays in planting and harvest as a result of an inability to effectively operate farming machinery, which can be strongly detrimental for crop production (Rosenzweig et al. 2002).

Rising Sea Level

Increases in sea level have been observed throughout the twentieth and early twenty-first centuries (Nicholls and Cazenave 2010). The major causes of this increase are thermal expansion of the oceans, and increases in the quantity of water in the oceans due to melting of major land ice structures (e.g. glaciers and ice caps/sheets), both of which are related to increases in global temperature (Nicholls and Cazenave 2010). Continued sea level rise is regarded as an inevitable consequence of climate change; a rise in sea level of up to 0.55 m under a low emissions scenario, and up to 0.98 m under a high emissions scenario, is predicted by 2100 (Stocker et al. 2013).

Sea-level rise will impact directly upon agriculture in low-lying regions globally due to flooding and salinization of groundwater (Gornall et al. 2010). However, the percentage of agricultural land that will be affected by sea-level rise during this century is likely to be relatively small (Dasgupta et al. 2009). Nevertheless, dramatic impacts are set to occur in some regions, e.g. the Vietnamese Mekong delta (Wassmann et al. 2004) and the Chinese Yangtze delta (Chen and Zong 1999).

Direct Effects of Carbon Dioxide and Ozone

As well as being the main drivers of anthropogenic climate change, greenhouse gases (GHGs) produced as a result of human activity can directly affect plant growth and development; carbon dioxide (CO_2) and ozone (O_3) are the most important factors in this regard. While there is variation amongst studies and debate in the literature about the extent of the effects of these gases, it is generally accepted that increases in CO_2 concentration likely to occur during the twenty-first century will exert a positive effect on crop productivity, while increases in O_3 concentration will exert a negative effect (Chen and Zong 1999).

The effect of elevated CO_2 concentration on crop performance has been the subject of considerable study; CO_2 is known to stimulate carbon assimilation in C_3 plant species, and to decrease stomatal conductance in both C_3 and C_4 species (Ainsworth and Rogers 2007). Both of these mechanisms can lead to increased crop productivity, however the effect of the latter is generally only detectable under water-limited conditions (Wang et al. 2008). Extensive studies conducted on plants grown in enclosures (e.g. controlled environment chambers, transparent field enclosures, or open-top chambers), have demonstrated large and highly significant gains in productivity associated with increased CO_2 concentration (Long et al. 2006). However, free-air concentration enrichment (FACE; Hendrey and Kimball 1994) experiments, which are designed to more accurately replicate growth conditions in future scenarios of elevated atmospheric CO_2 than enclosure studies, suggest substantially smaller effects than predicted by such enclosure studies (Long et al. 2006).

Air pollutants produced by human activity such as nitrogen oxides, carbon monoxide, and methane are ‘ozone precursors’, which react with hydroxyl radicals in the presence of UV to produce ground level (‘tropospheric’) O_3. Concentrations of tropospheric ozone vary substantially and are related to the local output of ozone precursors. In regions of high air pollution output, peak tropospheric O_3 concentration is much greater now than during pre-industrial times; furthermore there have been observable increases across most global regions, even far away from areas of high pollution output (Stevenson et al. 2013). As outlined by Wilkinson et al. (2012), O_3 can have detrimental effects on crops via several mechanisms, and current tropospheric levels of O_3 have been predicted to reduce global yields of some crops by up to 15 %. Effects of O_3 on crops include early induction of leaf senescence, visible injury to foliage and fruit, reduced carbon uptake and/or fixation, reduced carbon transportation, modulation of flowering, pollen sterility and ovule abortion, and (at least in some cases) reduced tolerance to abiotic stresses via effects on the control of stomatal regulation.

Crop Diseases and Pests

It has been estimated that diseases and pests cost global agriculture ~29 % of production (Oerke 2006). Changes in climate and atmospheric CO_2 concentration have the potential to influence the distribution, abundance, and aggressiveness of these diseases and pests (Luck et al. 2011). However, predicting the overall impact of climate change on crop losses associated with diseases and pests is challenging, in part because of the wide variety of pests and pathogens that may be differentially influenced. Additionally, the effects of climate and greenhouse gases on crops may affect the interaction of these crops with pests and pathogens. For example, several effects of elevated CO_2 concentration on plant growth and development may have indirect effects on crop disease, e.g. increases in leaf wax and epidermal thickness may provide enhanced defence against certain foliar pathogens, changes in canopy humidity may affect proliferation of some pathogens, and increased biomass accumulation may result in an increased reservoir for pathogen colonization (Luck et al. 2011). Recently, West et al. (2012) and Luck et al. (2011) have presented assessments of the effect of changes in climate and CO_2 concentration on diseases of major crops during the twenty-first century. They conclude that both increases and decreases of the incidence of specific diseases will occur, and that this is influenced by the lifestyle of the pathogen, the crop species being infected, and the geographic location of cultivation.

In addition to the ~12 % of crop production lost to disease, weeds and animal pests (primarily insects) are each responsible for losses of ~8 % (Oerke 2006). Currently there is limited data available regarding the potential impact of climate change on crop losses from these pests, nevertheless, there is potential for changes in climate and greenhouse gases to exert a substantial effect via several mechanisms. For example, changes in temperature may alter the distribution of weeds and animal pests (Ziska et al. 2011). Additionally, elevated temperatures will lead to increased reproduction of some weeds and animal pests, which may result in greater crop damage (DeLucia et al. 2012). Changes in nutrient profiles within plant tissues as a result of elevated CO_2 concentration may alter their attractiveness and/or nutritional value to animal pests (DeLucia et al. 2012). Furthermore, altered temperature and CO_2 and/or O_3 concentration may affect the competitiveness of weeds with crops; this is particularly likely where the crop and weed species use contrasting forms of photosynthesis that respond differently to these changes (Mahajan et al. 2012).

Due to the lack of certainty surrounding the impact of climate change on crop losses from pest and diseases, most models of climate change on crop productivity cannot incorporate these factors. This can be considered a major limitation for our ability to accurately predict future crop productivity scenarios (Gregory et al. 2009).

Pollinating Insects

The majority of total crop production globally comes from a small number of species that do not require animal pollinators, instead primarily relying on self-fertilization (e.g. wheat) or outcrossing via wind-pollination (e.g. maize and hybrid rice). Nevertheless, pollinator fertilization contributes substantially to agriculture; as outlined by (Aizen et al. 2009), 70 % of tropical crops, and 85 % of crops cultivated in Europe, benefit to some degree (from minor gains in yield, to complete dependence for reproduction) from biotic pollinators. Aizen et al. (2009) estimated that pollinating insects contribute 5 and 8 % of total crop production in the developed and developing world, respectively. Importantly, several nutrients that are critical for human well-being (e.g. vitamin C, lycopene, and folic acid) are obtained primarily from crops that rely on biotic pollinators (Eilers et al. 2011).

Changes in climate can directly affect the geographic distribution and lifecycle of both plants and pollinators. This has the potential to create spatial or temporal 'mismatches' between crops and their pollinators (Hegland et al. 2009), which could lead to decreased crop productivity (Memmott et al. 2007). There appears to be overlap in the response to temperature of phenology in many plants and pollinating insects, suggesting that these interactions might generally be tolerant of climate change; however, detailed understanding of this phenomenon is currently lacking (Hegland et al. 2009). With increasing demands on crop productivity, and given the importance of these interactions for the supply of certain essential nutrients, even relatively small negative impacts as a result of crop-pollinator mismatches could be problematic.

Strategies for Adaptation of Crop Production in Response to Climate Change

Adaptation has the potential to mitigate negative impacts of climate change on crop productivity. The use of adaptation for this purpose appears promising given that it has been of central importance to agriculture historically, allowing for the successful cultivation of the same or similar crops across a range of relatively diverse environments. Nevertheless, adaptation strategies will need to be tailored according to details of the location, climate change scenario, and the crop or crops being cultivated within each agricultural system that is being addressed (Rosenzweig and Tubiello 2007).

Traditional farm-level methods of crop adaptation include adjusting timing of planting, irrigation, fertilizer and pesticide applications, and the use of varieties that are more suited to the target cropping conditions. In some circumstances, these methods may be effective to offset potential yield decreases from changes in climate. For example, Tubiello et al. (2002) presented results of modelling for three US locations suggesting that for spring wheat, moving planting date forward would fully offset negative effects of temperature increases projected for this century. Their models also suggested that the selection of winter wheat cultivars with increased length of the grain filling period could partially compensate for yield losses associated with acceleration of development by increased temperature. However, traditional on-farm options that allow rapid adaptation with limited disruption will not be effective in all cases. Examples of potential limitations include the degree of phenological variation in available varieties, and the extent to which the cultivation period can be shifted without negative impacts due to other environmental factors (e.g. precipitation; soil moisture content) (Tubiello et al. 2002; Reilly et al. 2003).

Where farm-level adaptation strategies are not effective, more extreme, coordinated approaches might be. Breeding of elite cultivars for future climate scenarios is a possible option; however this can take more than a decade via traditional approaches. Genetic modification (GM) has the potential to facilitate the efficient production of highly adapted varieties (Varshney et al. 2012), however GM crops currently have low consumer acceptance and are subject to strong regulatory restrictions in most parts of the world. Perhaps the most extreme adaptation measure would be a shift in the area/s dedicated to crop cultivation within a given broad geographic region (Fischer et al. 2002). Clearly, implementation of such a strategy would be highly disruptive and require vast investment in planning and development of infrastructure.

The Impact of Crop Production on Climate Change, and Strategies to Mitigate This Impact

While crop production is at risk from anthropogenic climate change, it is itself a major contributor to climate change (Rosenzweig and Tubiello 2007). For example, Smith et al. (2007) reported that during 2005 agricultural production directly accounted for up to 12 % of total anthropogenic GHG emissions. Furthermore, a comparable scale of emission is indirectly driven by agriculture, via land clearing, production of fertilizers, and production and operation of farming machinery (Canadell et al. 2007). There is substantial potential for mitigation of GHG emissions from crop production (Smith and Olesen 2010). The development and/or adoption of strategies to reduce emissions from agricultural cultivation can clearly contribute to mitigation efforts. Additionally, practices that lead to increased sequestration of carbon in soil can mitigate atmospheric CO_2 accumulation (Lal 2004).

N_2O produced via the application of nitrogenous fertilizers accounts for over a third of GHG emissions from crop production (Burney et al. 2010). Maintaining adequate soil nitrogen concentration through fertilization is essential to sustain high crop yields, however, excessive fertilizer application is commonplace globally and much of the excess nitrogen is lost to the atmosphere as N_2O (Conyers et al. 2013). Thus, ‘nutrient budgeting’ (Conyers et al. 2013) (i.e. closely matching fertilizer inputs to crop requirements) can be highly effective to mitigate N_2O emissions; precision agriculture (i.e., broadly, the use of monitoring technologies to identify and respond to inter and intra-field variation to improve the efficiency of crop production) has great potential for this purpose (Fulton et al. 2013).

Approximately 9 % of anthropogenic methane (CH_4) emissions occur as a result of rice cultivation, via the metabolism of CH_4 producing microorganisms inhabiting flooded rice paddies. If farming practices remain unchanged, this is predicted to increase under future elevated atmospheric CO_2 concentration (van Groenigen et al. 2013). Options exist for strong mitigation of the contribution of rice cultivation to CH_4 production. Yan et al. (2009) estimated that global adoption of the draining of rice paddies at least once during the cropping cycle, in conjunction with the application of rice straw (widely used as a fertilizer) to drained rather than flooded paddies, could reduce CH_4 production by ~32 %. The use of biological and/or organic amendments during rice cultivation also has potential to reduce CH_4 production (e.g. Ali et al. 2008).

There are several viable strategies to increase carbon sequestration via crop production. Reduced or no tillage practices, crop rotations with forage crops, and ley-arable practices (Vertès et al. 2007) are effective in returning carbon to soil (Rosenzweig and Tubiello 2007), and soil amendments can also increase carbon storage (Mandal et al. 2007). Additionally, agroforestry has shown great potential for carbon sequestration (Ramachandran Nair et al. 2009). Rosenzweig and Tubiello (2007) estimated that by implementation of all available agricultural strategies for carbon sequestration, up to 30 gigatonnes of carbon could be stored over the next 40 years, equivalent to ~60 % of the carbon lost from agricultural soils historically.

As stated above, land clearing to facilitate crop production contributes substantially to greenhouse gas emissions; therefore, minimizing the expansion of agricultural land represents a further mitigation strategy. Dramatic yield gains in staple crops have allowed production to increase by 135 % with an expansion of agricultural land of only 27 % since 1961 (Burney et al. 2010). Achieving productivity gains primarily via intensification rather than expansion during this period has been estimated to avoid emissions equivalent to 161 gigatonnes of carbon (Burney et al. 2010). This is approximately equal to the total volume of carbon lost to the atmosphere by both land conversion and subsequent agricultural production during the last century (Rosenzweig and Tubiello 2007). Thus, in addition to its critical importance to provide for the calorie demands of a rapidly expanding human population (Ray et al. 2013), improving yield potential of elite crop varieties represents a highly effective strategy to minimize the impact of agriculture on GHG emissions.

Conclusions

It is clear that recent human activity has affected climate, with drastic increases in greenhouse gas emissions leading to a substantial rise in mean global surface temperature during the last century. The best currently available models predict continuing changes in climate over the next 100 years. There is substantial evidence that climate change and the increase in atmospheric concentration of greenhouse gases are already affecting the performance of our staple crops. With rapidly increasing calorie demands from an expanding human population, assessing the future impact of these phenomena on crop productivity is of the utmost importance; however, this is extremely challenging due to the multitude of complex factors at play. Current models suggest that over the next 100 years increases in CO_2 concentration will exert a positive influence on productivity of our major crops, while increases in temperature and O_3 concentration will exert a negative influence. Future models will be refined in their ability to account for these factors, and should begin to meaningfully incorporate the influence of other important phenomena (e.g. the effects of climate change on crop pests and diseases). This will improve our ability to quantify the risk that anthropogenic climate change poses to crop productivity and future food security. While there is substantial uncertainty surrounding the short-to-medium term net influence of GHG emissions and climate change on crop productivity, it appears clear that continued increase in emissions, temperature, and extreme weather events will ultimately be strongly detrimental. Thus, the reduction of GHG emissions appears crucial for future food security, and the mitigation of emissions from crop production itself can assist this process.

References

Ainsworth EA, Rogers A (2007) The response of photosynthesis and stomatal conductance to rising [CO_2]: mechanisms and environmental interactions. Plant Cell Environ 30(3):258–270

Aizen MA, Garibaldi LA, Cunningham SA, Klein AM (2009) How much does agriculture depend on pollinators? Lessons from long-term trends in crop production. Ann Bot 103:1579–1588

Ali MA, Oh JH, Kim PJ (2008) Evaluation of silicate iron slag amendment on reducing methane emission from flood water rice farming. Agr Ecosyst Environ 128(1):21–26

Ashraf M (1999) Interactive effects of nitrate and long-term waterlogging on growth, water relations, and gaseous exchange properties of maize (Zea mays L.). Plant Sci 144(1):35–43

Barnabás B, Jäger K, Fehér A (2008) The effect of drought and heat stress on reproductive processes in cereals. Plant Cell Environ 31(1):11–38

Burney JA, Davis SJ, Lobell DB (2010) Greenhouse gas mitigation by agricultural intensification. Proc Natl Acad Sci 107(26):12052–12057

Canadell JG, Le Quéré C, Raupach MR, Field CB, Buitenhuis ET, Ciais P et al (2007) Contributions to accelerating atmospheric CO2 growth from economic activity, carbon intensity, and efficiency of natural sinks. Proc Natl Acad Sci 104(47):18866–18870

Carnesale A, Chameides W (2011) America's climate choices. NRC/NAS USA Committee on America's Climate Choices. http://download nap edu/cart/delivercgi

Cassidy ES, West PC, Gerber JS, Foley JA (2013) Redefining agricultural yields: from tonnes to people nourished per hectare. Environ Res Lett 8(3):034015

Cassman KG, Dobermann A, Walters DT, Yang H (2003) Meeting cereal demand while protecting natural resources and improving environmental quality. Annu Rev Environ Resour 28(1):315–358

Challinor A, Watson J, Lobell D, Howden S, Smith D, Chhetri N (2014) A meta-analysis of crop yield under climate change and adaptation. Nat Clim Change 4(4):287–291

Chen X, Zong Y (1999) Major impacts of sea-level rise on agriculture in the Yangtze delta area around Shanghai. Appl Geogr 19(1):69–84

Conyers M, Bell M, Wilhelm N, Bell R, Norton R, Walker C (2013) Making better fertiliser decisions for cropping systems in Australia (BFDC): knowledge gaps and lessons learnt. Crop Pasture Sci 64(5):539–547

Coumou D, Rahmstorf S (2012) A decade of weather extremes. Nat Clim Change 2(7):491–496

Dasgupta S, Laplante B, Meisner C, Wheeler D, Yan J (2009) The impact of sea level rise on developing countries: a comparative analysis. Clim Change 93(3-4):379–388

DeLucia EH, Nabity PD, Zavala JA, Berenbaum MR (2012) Climate change: resetting plant-insect interactions. Plant Physiol 160(4):1677–1685

Deryng D, Conway D, Ramankutty N, Price J, Warren R (2014) Global crop yield response to extreme heat stress under multiple climate change futures. Environ Res Lett 9(3):034011

Eilers EJ, Kremen C, Greenleaf SS, Garber AK, Klein A-M (2011) Contribution of pollinator-mediated crops to nutrients in the human food supply. PLoS One 6(6), e21363

Fischer G, Shah M, Van Velthuizen H (2002) Climate change and agricultural vulnerability. IIASA, Laxenburg

Fulton JP, Shearer SA, Higgins SF, McDonald TP (2013) A method to generate and use as-applied surfaces to evaluate variable-rate fertilizer applications. Precis Agric 14(2):184–200

Gornall J, Betts R, Burke E, Clark R, Camp J, Willett K et al (2010) Implications of climate change for agricultural productivity in the early twenty-first century. Philos Trans R Soc B Biol Sci 365(1554):2973–2989

Gregory PJ, Johnson SN, Newton AC, Ingram JS (2009) Integrating pests and pathogens into the climate change/food security debate. J Exp Bot 60(10):2827–2838

Hegland SJ, Nielsen A, Lázaro A, Bjerknes AL, Totland Ø (2009) How does climate warming affect plant-pollinator interactions? Ecol Lett 12(2):184–195

Hendrey G, Kimball B (1994) The FACE program. Agr Forest Meteorol 70(1):3–14

Keckhut P, Cagnazzo C, Chanin M-L, Claud C, Hauchecorne A (2005) The 11-year solar-cycle effects on the temperature in the upper-stratosphere and mesosphere: part I—assessment of observations. J Atmos Sol-Terr Phys 67(11):940–947

Kozdrój J, van Elsas JD (2000) Response of the bacterial community to root exudates in soil polluted with heavy metals assessed by molecular and cultural approaches. Soil Biol Biochem 32(10):1405–1417

Lal R (2004) Soil carbon sequestration impacts on global climate change and food security. Science 304(5677):1623–1627

Leinonen I, Jones HG (2004) Combining thermal and visible imagery for estimating canopy temperature and identifying plant stress. J Exp Bot 55(401):1423–1431

Lobell DB, Gourdji SM (2012) The influence of climate change on global crop productivity. Plant Physiol 160(4):1686–1697

Long SP, Ainsworth EA, Leakey AD, Nösberger J, Ort DR (2006) Food for thought: lower-than-expected crop yield stimulation with rising CO2 concentrations. Science 312(5782):1918–1921

Luck J, Spackman M, Freeman A, Griffiths W, Finlay K, Chakraborty S (2011) Climate change and diseases of food crops. Plant Pathol 60(1):113–121

Mahajan G, Singh S, Chauhan BS (2012) Impact of climate change on weeds in the rice-wheat cropping system. Curr Sci 102(9):1254–1255

Mandal B, Majumder B, Bandyopadhyay P, Hazra G, Gangopadhyay A, Samantaray R et al (2007) The potential of cropping systems and soil amendments for carbon sequestration in soils under long-term experiments in subtropical India. Glob Chang Biol 13(2):357–369

Memmott J, Craze PG, Waser NM, Price MV (2007) Global warming and the disruption of plant–pollinator interactions. Ecol Lett 10(8):710–717

Nicholls RJ, Cazenave A (2010) Sea-level rise and its impact on coastal zones. Science 328(5985):1517–1520

Oerke EC (2006) Crop losses to pests. J Agric Sci 144(01):31–43

Parry ML (2007) Climate change 2007: impacts, adaptation and vulnerability: contribution of Working Group II to the fourth assessment report of the Intergovernmental Panel on Climate Change. Cambridge University Press, Cambridge

Polley HW (2002) Implications of atmospheric and climatic change for crop yield and water use efficiency. Crop Sci 42(1):131–140

Porter JR, Semenov MA (2005) Crop responses to climatic variation. Philos Trans R Soc B Biol Sci 360(1463):2021–2035

Prescott-Allen R, Prescott-Allen C (1990) How many plants feed the world? Conserv Biol 4(4): 365–374

Ramachandran Nair P, Mohan Kumar B, Nair VD (2009) Agroforestry as a strategy for carbon sequestration. J Plant Nutr Soil Sci 172(1):10–23

Ramankutty N, Foley JA, Norman J, McSweeney K (2002) The global distribution of cultivable lands: current patterns and sensitivity to possible climate change. Glob Ecol Biogeogr 11(5): 377–392

Ray DK, Mueller ND, West PC, Foley JA (2013) Yield trends are insufficient to double global crop production by 2050. PLoS One 8(6), e66428

Reilly J, Tubiello F, McCarl B, Abler D, Darwin R, Fuglie K et al (2003) US agriculture and climate change: new results. Clim Change 57(1-2):43–67

Rosenzweig C, Neofotis P (2013) Detection and attribution of anthropogenic climate change impacts. Wiley Interdiscip Rev Clim Change 4(2):121–150

Rosenzweig C, Tubiello FN (2007) Adaptation and mitigation strategies in agriculture: an analysis of potential synergies. Mitig Adapt Strat Glob Chang 12(5):855–873

Rosenzweig C, Tubiello FN, Goldberg R, Mills E, Bloomfield J (2002) Increased crop damage in the US from excess precipitation under climate change. Glob Environ Chang 12(3):197–202

Schmidhuber J, Tubiello FN (2007) Global food security under climate change. Proc Natl Acad Sci 104(50):19703–19708

Smith P, Olesen JE (2010) Synergies between the mitigation of, and adaptation to, climate change in agriculture. J Agric Sci 148(5):543–552

Smith P, Martino D, Cai Z, Gwary D, Janzen H, Kumar P et al (2007) Policy and technological constraints to implementation of greenhouse gas mitigation options in agriculture. Agr Ecosyst Environ 118(1):6–28

Spiegel DS, Raymond SN, Dressing CD, Scharf CA, Mitchell JL (2010) Generalized Milankovitch cycles and long-term climatic habitability. Astrophys J 721(2):1295–1307

Stevenson D, Young P, Naik V, Lamarque J-F, Shindell DT, Voulgarakis A et al (2013) Tropospheric ozone changes, radiative forcing and attribution to emissions in the Atmospheric Chemistry and Climate Model Intercomparison Project (ACCMIP). Atmos Chem Phys 13(6):3063–3085

Stocker T, Qin D, Plattner G, Tignor M, Allen S, Boschung J et al (2013) IPCC, 2013: climate change 2013: the physical science basis. Contribution of working group I to the fifth assessment report of the intergovernmental panel on climate change

Sutton RT, Dong B, Gregory JM (2007) Land/sea warming ratio in response to climate change: IPCC AR4 model results and comparison with observations. Geophys Res Lett 34(2), L02701

Teixeira EI, Fischer G, van Velthuizen H, Walter C, Ewert F (2013) Global hot-spots of heat stress on agricultural crops due to climate change. Agr Forest Meteorol 170:206–215

Trenberth KE (2012) Framing the way to relate climate extremes to climate change. Clim Change 115:283–290

Tubiello F, Rosenzweig C, Goldberg R, Jagtap S, Jones J (2002) Effects of climate change on US crop production: simulation results using two different GCM scenarios. Part I: wheat, potato, maize, and citrus. Climate Res 20(3):259–270

van Groenigen KJ, van Kessel C, Hungate BA (2013) Increased greenhouse-gas intensity of rice production under future atmospheric conditions. Nat Clim Change 3(3):288–291

Varshney RK, Ribaut J-M, Buckler ES, Tuberosa R, Rafalski JA, Langridge P (2012) Can genomics boost productivity of orphan crops? Nat Biotechnol 30(12):1172–1176

Vertès F, Hatch D, Velthof G, Taube F, Laurent F, Loiseau P et al (2007) Short-term and cumulative effects of grassland cultivation on nitrogen and carbon cycling in ley-arable rotations. Permanent and temporary grassland: plant, environment and economy. Proceedings of the 14th symposium of the European Grassland Federation, Ghent, Belgium, 3–5 September 2007; Belgian Society for Grassland and Forage Crops

Wang D, Heckathorn SA, Barua D, Joshi P, Hamilton EW, LaCroix JJ (2008) Effects of elevated CO2 on the tolerance of photosynthesis to acute heat stress in C3, C4, and CAM species. Am J Bot 95(2):165–176

Wassmann R, Hien NX, Hoanh CT, Tuong TP (2004) Sea level rise affecting the Vietnamese Mekong Delta: water elevation in the flood season and implications for rice production. Clim Change 66(1-2):89–107

West JS, Townsend JA, Stevens M, Fitt BD (2012) Comparative biology of different plant pathogens to estimate effects of climate change on crop diseases in Europe. Eur J Plant Pathol 133(1):315–331

Wheeler TR, Craufurd PQ, Ellis RH, Porter JR, Prasad PV (2000) Temperature variability and the yield of annual crops. Agr Ecosyst Environ 82(1):159–167

Wilkinson S, Mills G, Illidge R, Davies WJ (2012) How is ozone pollution reducing our food supply? J Exp Bot 63(2):527–536

Wilson PR (2003) Solar and stellar activity cycles. Cambridge University Press, Cambridge

Yamori W, Hikosaka K, Way DA (2014) Temperature response of photosynthesis in C3, C4, and CAM plants: temperature acclimation and temperature adaptation. Photosynth Res 119(1-2):101–117

Yan X, Akiyama H, Yagi K, Akimoto H (2009) Global estimations of the inventory and mitigation potential of methane emissions from rice cultivation conducted using the 2006 Intergovernmental Panel on Climate Change guidelines. Global Biogeochem Cycles 23(2):15 pages

Yin X, Kropff MJ, McLaren G, Visperas RM (1995) A nonlinear model for crop development as a function of temperature. Agr Forest Meteorol 77(1):1–16

Ziska LH, Blumenthal DM, Runion GB, Hunt ER Jr, Diaz-Soltero H (2011) Invasive species and climate change: an agronomic perspective. Clim Change 105(1-2):13–42

The Impacts of Extreme Climatic Events on Wild Plant Populations

Robert C. Godfree, Lyndsey M. Vivian, and Jennifer C. Pierson

Introduction

Anthropogenic changes to the earth's climate over the past few decades have spurred a growing level of scientific interest in the impact of rapid climate change on global ecosystems. To date, the majority of studies have focused on quantifying the impact of mean climatic trends, such as warming temperatures or declining frost period, on the phenology (Menzel et al. 2006), behaviour (Parmesan et al. 2000) and distribution of individual species and biotic assemblages (Chen et al. 2011). There is now unequivocal evidence that significant shifts in both species and ecosystem attributes in response to broader climatic trends are occurring worldwide (Walther et al. 2002). So ubiquitous are these changes that they constitute a major threat to the long-term conservation of global biodiversity, and may even alter or disrupt ecosystem processes that underpin the functioning of earth's biosphere itself (Cramer et al. 2001; Hannah et al. 2002; Gerten 2013; Doblas-Miranda et al. 2014).

Arguably, much less effort has been invested in understanding the roles that extreme or unusual climatic or weather events (hereafter extreme climatic events, ECEs) play in driving population and ecosystem change. This is surprising, because climatic extremes have in the past resulted in profound human societal change, and apparently even the collapse of prehistorical cultures and the agro-ecosystems on

R.C. Godfree, Ph.D. (✉)
National Research Collections Australia, Commonwealth Scientific and Industrial Research Organisation, Clunies Ross Street, Canberra, ACT 2601, Australia
e-mail: Robert.Godfree@csiro.au

L.M. Vivian, Ph.D.
National Research Collections Australia, CSIRO, Canberra, ACT Australia

J.C. Pierson, Ph.D.
Fenner School of Environment and Society, The Australian National University, Acton, ACT Australia

D. Edwards, J. Batley (eds.), *Plant Genomics and Climate Change*,
DOI 10.1007/978-1-4939-3536-9_2

Fig. 1 A stone pueblo constructed by Ancestral Puebloan (Anasazi) people at Hovenweep National Monument, Utah, USA. Hovenweep people farmed maize and other domesticated crops on surrounding terraced fields despite the semi-arid climate. As with many other Anasazi sites in the region, Hovenweep was abandoned in the late thirteenth century CE, probably at least partly in response to extreme, multi-decadal drought. [Courtesy of Holly Godfree]

which they depended. For example, multidecadal drought during the middle-twelfth and late-thirteenth-century CE (Common Era) is thought to have led to the dislocation of maize-dependent Anasazi (Ancestral Puebloan) people in the American southwest (Benson et al. 2007; Fig. 1), while the decline of the Classic Mayan civilisation coincided with a period of recurring drought between ca. 770 and 1100 CE (Hodell et al. 2005). Contemporary events, such as Australia's 1997–2009 'Millennium Drought' (Murphy and Timbal 2008; Heberger 2012) and the 2003 European heatwave (Stott et al. 2004; Ciais et al. 2005) clearly demonstrate that extreme climatic or weather conditions can be potent agents of change in anthropogenic and natural systems alike.

One reason why few studies have attempted to untangle the complex linkages that exist between climate variability, climatic trends and biotic change is that it is logistically difficult to quantify the impact of statistically improbable climatic or meteorological events on single populations, let alone entire ecosystems. Simulation of extreme climatic events and climate change under realistic field conditions, while desirable, is usually expensive, time consuming and often technically challenging (e.g., Hovenden et al. 2008; Smith 2011; Dieleman et al. 2012; Godfree et al. 2013), and so most existing studies report on the impacts of naturally occurring ECEs on plant or animal populations or communities. Nonetheless, combined with data collected using remote sensing techniques (Zhao and Running 2010) and controlled studies that artificially induce extreme physiological stress in experimental populations (e.g., Musil et al. 2005; Marchand et al. 2006), the body of work that does exist demonstrates that relatively brief or protracted periods of unusual climate or weather can impact on biota at scales ranging from the individual organism through to entire ecosystems. In a world facing significant warming in the decades ahead, the possibility that ECEs will be fundamental drivers of ecosystem change cannot be ignored.

The objective of this chapter is to provide an overview of the nature of extreme climatic or climate-linked events and their impacts on the demography of wild plant populations. While a complete review of this subject is beyond the scope of this chapter, we focus on drawing out some of the main lessons that have been learned from the past and contemporary study of ECEs, and apply them to understanding the potential demographic and evolutionary roles that extreme events may play in the future within the context of anthropogenic climate change. While we provide a global perspective on the subject, we focus especially on specific case studies involving Australian vegetation, since the principles learned from these events are applicable to extreme events more generally. The Australian environment is highly suited to the study of ECEs, where they have left an indelible mark on both human societies and on continental biota alike. We first discuss the definition and statistical characterisation of ECEs, a source of much confusion in the literature.

Characterisation of Extreme Climatic Events

One significant complication associated with the study of climate extremes is that there is no universally accepted definition of what actually constitutes an extreme event. Many studies investigating the role of extreme events on plant populations do not quantify either the statistical probability of the climatic conditions observed during the study nor the magnitude of their effects on biota relative to what is considered background or normal variation. Hence, the notion of "extremeness" is often vague or ill-defined and provides little basis for extrapolation to other systems or events. Many simply define drought severity in terms of departure from mean rainfall, providing little, if any, data on inter-annual rainfall variability (e.g., Condit et al. 1995), despite the fact that a rigorous statistical theory of extreme events now exists that greatly facilitates the quantification of such events (Katz et al. 2005). More recently, it has been argued that one should consider both the probability of occurrence of both the climatic event and the ecosystem response, and that an 'extreme event' must involve both a rare climatic event and an unusually large ecosystem response (e.g., Smith 2011). To complicate matters further, the terminology used to describe climate and weather extremes in the scientific literature varies widely, often involving a mix of terms that describe attributes such as rarity, severity, magnitude of impact, and event complexity (Stephenson 2008).

To avoid confusion, we adopt the Intergovernmental Panel in Climate Change (IPCC) definition of what constitutes an extreme climate event (ECE), and provide (where possible) the statistical probability of the event using indices based on either a standard climatic reference period or a given temperature or precipitation threshold. The IPCC (IPCC 2012; Section 3.1.2) describes an 'extreme climate or weather event' in statistical terms: "the occurrence of a value of a weather or climate variable above (or below) a threshold value near the upper (or lower) ends of the range of observed values of the variable" (Section 3.1.2, p. 116). By following the IPCC convention we explicitly separate the concept of the rarity of a climatic event from the population- or ecosystem-level impact of the event. Indeed some extreme events appear to have little or no impact on vegetation, and so this distinction is in itself an interesting area for inquiry.

Probably the most commonly used indices quantify climatic extremes in terms of percentiles or quantiles (especially deciles) relative to a baseline distribution, often the World Meteorological Organisation's 1961–1990 normal reference period (this is to be replaced in 2020 with a 1991–2020 reference period). For example, a tenth percentile (= first decile) annual rainfall total of 1 year duration has a 10 % chance of occurring in a given year, with an average return interval of 10 years. Others are based on frequency or duration over a specific threshold (e.g., the number of hot days exceeding 40 °C), or, in the case of chronic or multiple extremes, more complex definitions. Formal statistical distribution fitting techniques may also be used to estimate the probability of unusual events occurring in a given timeframe, in which case the choice of distribution is critically important, including from an environmental policy standpoint (Pindyck 2011).

Some of the key statistical aspects of ECEs, and the complexities associated with quantifying chronic events, can be illustrated by examining variation in total annual rainfall between 1889 and 2012 at Barraba Post Office (S 30.3781°, E 150.6096°) in northern NSW, Australia (see Fig. 2). Barraba has a nearly continuous rainfall record over this period, and shows a pattern of variability that is broadly representative of much of northern NSW. Over the entire record rainfall averaged 692 mm, with 5th and 95th percentile rainfall totals (here denoted P_5 and P_{95}) of 431 and 957 mm respectively (Fig. 3a), with high inter-annual variability (Fig. 3a). However, rainfall behaviour and the nature of extreme events have changed considerably over the past century: prior to 1950 there was a protracted period of lower (average $=\mu=656$ mm) but less variable rainfall ($P_5=450$ mm, $P_{95}=887$ mm), with 1896–1946 being especially dry ($\mu=621$ mm, $P_5=442$ mm, $P_{95}=834$ mm). The 1896–1902 'Federation Drought' and the 1937–1946 'World War II Droughts' which devastated agricultural communities and rangelands across SE Australia (McKeon et al. 2004) both occurred during this period.

Since 1949, and especially during the 1961–1990 reference period, rainfall has been higher ($\mu=726$ mm) but generally more variable ($P_5=397$ mm, $P_{95}=1007$ mm). Five of the seven driest years have occurred since 1949, including all 4 years with <400 mm rainfall (despite the general absence of multi-year droughts), as have 6 of the 7 wettest years. Differences in the statistical behaviour of rainfall during these two periods can be clearly seen in Fig. 3c, d, where annual totals have been fit

Fig. 2 Sites and events referred to in the text. (*A*) Barraba, NSW. (*B*) General location of *Eucalyptus marginata* (Jarrah) forest in south-western Western Australia. (*C*) *Austrostipa aristiglumis* grassland near Wyalong, NSW. (*D*) Frost affected rainforest vegetation in the Atherton Tablelands region. (*E*) Mediterranean mallee-heathland community affected by extreme heat. (*F*) Drought and insect attack in *Eucalyptus* spp. near Canberra, ACT. (*G*) Degraded semi-arid rangeland near Carinda, NSW. (*H*) Barmah-Millewa Forest

with a 3-parameter Weibull function. In contrast to the near-normally distributed rainfall of 1950–2012, the distribution for the period 1889–1949 is both narrower and skewed towards lower annual totals. The prolonged nature of rainfall deficiencies in 1896–1949 is effectively shown in the associated residual mass diagram (Fig. 3b), which displays cumulative annual rainfall deviations from the 1889–2012 mean. During this period accumulative rainfall deficiencies totalled 3618 mm, or roughly five times the average annual rainfall. Since 1949 rainfall surpluses have dominated.

Delving further, we can also see that the seasonal timing of rainfall at Barraba has also changed since 1889. The most significant changes have been a 45 % increase in summer (December–February) rainfall (Fig. 4a), which has resulted in a significant reduction in the frequency and severity of dry summers in recent decades, and a weak decline in autumn (March–May) rainfall (Fig. 4b), with only one even modestly wet autumn (top 25th percentile of rainfall totals) having been recorded in the

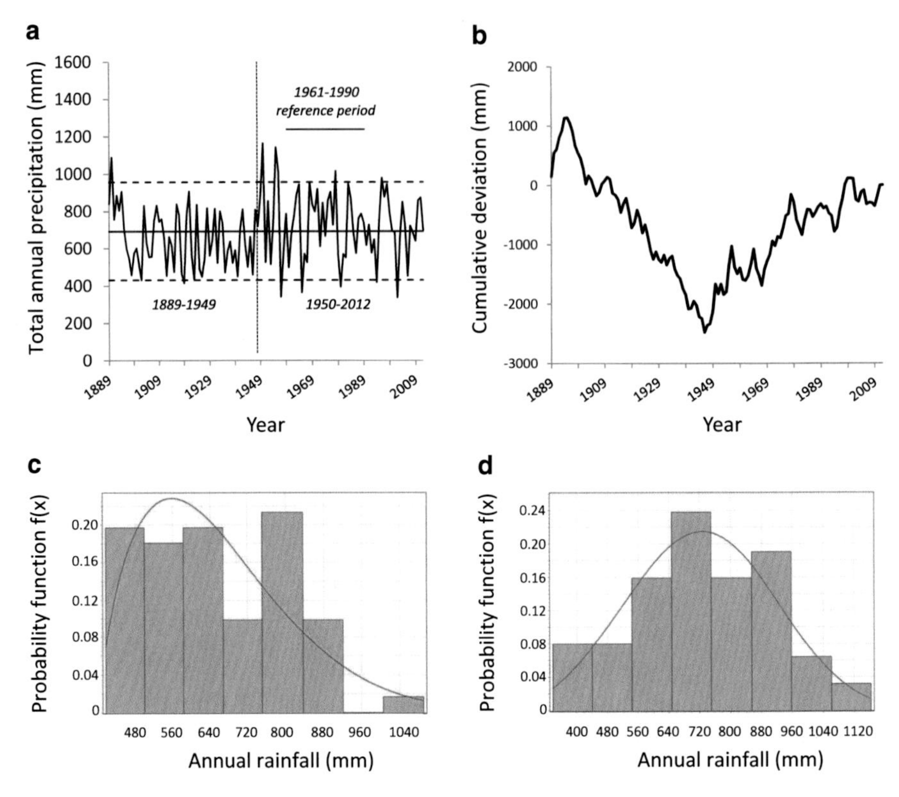

Fig. 3 Rainfall data for Barraba Post Office, northern NSW, Australia. (**a**) Total annual precipitation 1889–2012, showing the 1961–1990 reference period. (**b**) Cumulative residual mass diagram for 1889–2012 data, showing the long period of below average precipitation between 1896 and 1946 and the wetter conditions that followed. (**c**) Annual rainfall for 1889–1949 fit with a three - parameter Weibull function ($\alpha = 1.6$, $\beta = 282.9$ and $\gamma = 402.4$). (**d**) Annual rainfall for 1950–2012 fit with a three-parameter Weibull function ($\alpha = 3.4$, $\beta = 626.2$ and $\gamma = 163.7$)

past 21 years (Fig. 4b). A similar pattern of autumn rainfall decline has been observed across SE Australia in recent decades, especially during the 1997–2009 Millennium Drought. In a region of Australia which relies on autumn rain to establish crops, this change has significantly affected agricultural yield (Pook et al. 2009).

Collectively, these data illustrate several important features of extreme climate events and climate change. First, statistical theory indicates that the frequency of extreme events is more influenced by changes in variability than means (Katz and Brown 1992), but as the Barraba rainfall data show, both can change simultaneously, producing complex behaviour of climate extremes. Second, the statistical distribution of a climate variable, and hence estimates of the rarity of a given extreme event, can depend on the specific reference period considered. For example, the Weibull distribution model shown in Fig. 3c, based on data from 1889 to 1949, predicts a lower probability of recording a year with below 420 mm (1 %) than both

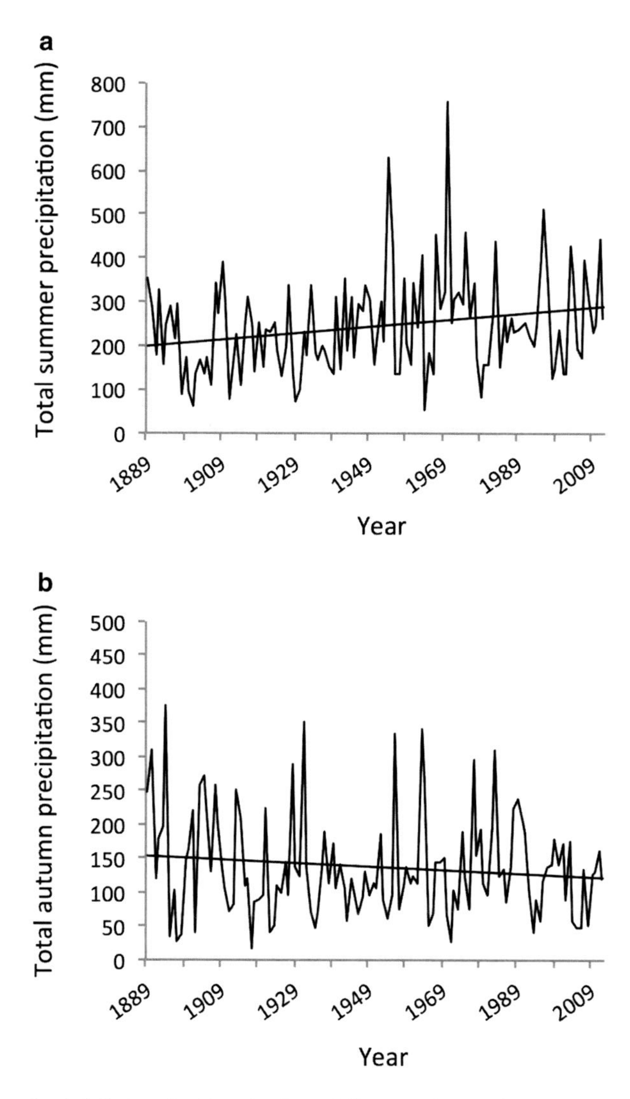

Fig. 4 Seasonal rainfall data for Barraba Post Office, northern NSW, Australia. (**a**) Summer (December–February). (**b**) Autumn (March–May)

the 1950–2012 (Fig. 3d; 5 %) and 1961–1990 reference periods (4 %). In fact, five such years have been recorded since 1889, and four since 1950. Changes can also occur in the seasonality or timing of ECEs, which may have important implications for plant adaptation. Finally, a range of statistical criteria and timeframes may be needed to adequately quantify a given ECE, especially one such as drought which may last for many years. For example, while many of the individual driest years at Barraba have occurred since 1950, the longest runs of below average rainfall years (up to seven) all occurred before 1950.

We now turn our attention to the impact of such events on wild plant populations. A review of the entire body of research relating to ECEs is well beyond the scope of this chapter, and so we focus on the phytosociological impacts of those events that drive rapid demographic changes in plant populations. Such events are of special significance given the emerging evidence for increased mortality and vegetation change in ecosystems globally in response to anthropogenic climate change.

ECEs and Population Demography

Mortality as a Key Driver of Demographic Change

By far the most well studied effects of ECEs on plant populations are those that involve the death of individuals as a direct result of water deficiencies and extremes of heat and cold. Since a comprehensive review of all such studies is beyond the scope of this chapter, the intent here is to draw key lessons from the literature.

Perhaps the most obvious conclusion to be drawn from an appraisal of the ECE literature is that most, if not all ecosystems globally can experience rapid demographic change in response to rare climatic or climate-related events, probably even in the absence of underlying climate change. Extensive mortality of plant populations directly resulting from drought stress has been observed in arid, semi-arid, temperate, subtropical and tropical forests and woodlands, arid, semi-arid and alpine grasslands and shrublands, to name but a few ecosystems. Negative effects of high temperatures, along or in combination with drought, and extremes of cold have also been reported in a great range of biomes. If we broaden our scope to include the effects of severe storms, snowfalls and hydrological events associated with extreme weather, probably few ecosystems are left untouched by extreme events at some point.

One of the earliest, yet still most comprehensive, accounts of plant population mortality and change in response to an ECE was provided by Albertson and Weaver (1944, 1945), who quantified the impacts of the 1930s "Great Drought" (Woodhouse and Overpeck 1998) on prairie and woodland ecosystems of the mid-western United States. Apparently caused by unusual tropical ocean temperatures coupled with land surface-atmosphere feedbacks (Schubert et al. 2004), this period of chronically dry conditions (punctuated by shorter periods of record low rainfall), in combination with record high temperatures and poor land management, had a catastrophic impact on agricultural and natural ecosystems alike. As vividly captured in John Steinbeck's novel *The Grapes of Wrath*, the development of the infamous "Dust Bowl" areas of the southern Great Plains ultimately became one of the greatest catastrophes in recent North American history. During this drought, widespread desiccation and death of woody and herbaceous vegetation occurred across the Great Plains, with post-climax vegetation along the western woodland margin and prairie grassland being generally the most severely affected.

Some of the most dramatic recent examples of plant mortality in response to ECEs have come from forests and woodlands, which seem to be highly susceptible to rapid change or even collapse once certain physiological thresholds are exceeded. The recent canopy collapse observed in Mediterranean-type *Eucalyptus marginata* (Jarrah)-dominated forest in south-western Australia (Fig. 2), for instance, appears to have been caused by record low rainfall and warm temperatures (culminating in a prolonged heatwave in February 2011), possibly along with declining groundwater levels (Matusick et al. 2013). Contemporary changes in other species including *Eucalyptus gomphocephala* (Matusick et al. 2012) and *Eucalyptus wandoo* (Brouwers et al. 2012) suggest that deteriorating water balance is a phenomenon increasingly affecting much of south-western Australia. This region is highly sensitive to changes in atmospheric circulation that result in more southerly tracking of the mid latitude fronts that provide much of its precipitation, synoptic conditions which are likely to become more prevalent under global warming (Hope 2006).

Case studies such as this reinforce the view that forests and other vegetation in many areas are likely to be placed at risk by an increase in the frequency and severity of extreme drought and heat events under future climate change (Allen et al. 2010; Anderlegg et al. 2013). Critically, they also show that demographic changes associated with climate extremes are often rapid, can occur over large spatial scales, and can drive significant structural and compositional change across multiple plant communities. As we discuss below, such patterns will affect the ability of plant populations and communities to adapt to climate change, which has strong implications for biodiversity conservation on a global scale.

Nonlinearity of ECE Impacts

During the 1930s Great Drought in the mid-west US, extensive tree mortality was not only observed during the exceptionally dry years of 1934 and 1936, but also in 1939 when rainfall deficiencies were less extreme. Indeed, some populations of trees that had survived previous dry years succumbed only during 1939. As noted by Albertson and Weaver (1945), this reflects the ongoing development of soil moisture deficits over the course of the drought, the effects of repeated tissue desiccation and injury, and more complex land surface-pathogen-plant interactions. This illustrates one of the more interesting aspects of ECEs: that the relationship between the demographic response of plant populations and increasing severity of climatic extremes, and indeed abiotic stress in general, is usually strongly nonlinear. In the most striking cases, changes are minimal, often for protracted periods, until a specific threshold is reached, after which demographic change becomes extremely rapid.

Consider the behaviour of a population of the perennial tussock grass species *Austrostipa aristiglumis* in semi-arid grassland in central NSW, Australia (see Godfree et al. 2011; Fig. 2) when subjected to an exceptionally severe drought. Between 2001 and 2009 the study site was affected by drought of varying intensity,

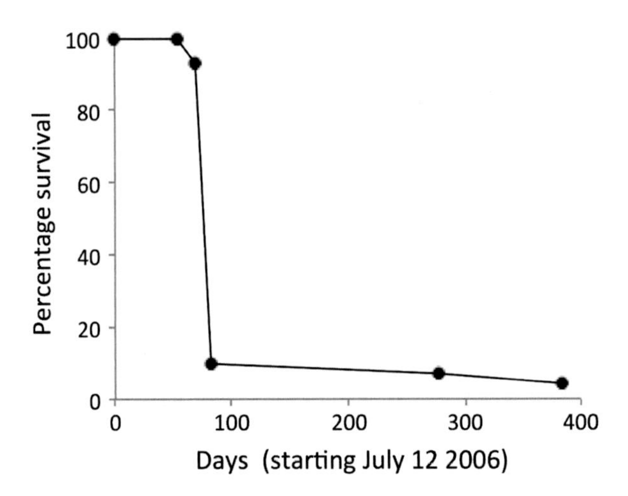

Fig. 5 Survival of adult *Austrostipa aristiglumis* during 2006–2007 at the Wyalong study site. The vast majority of mortality occurred in a 2-week period between late October and early November 2006

with 2006 being the driest year on record (181 mm; annual average = 480 mm). Between January and May 2006 only 28 mm of rain fell, leaving the soil profile bereft of plant-available water, but some recharge of the upper profile occurred in June and July. To this point, virtually no mortality of *A. aristiglumis* plants was observed. However, dry conditions re-established in August 2006, and virtually all adult plants underwent complete senescence of above and below-ground tissue within a 2-week period in late October 2006 (Figs. 5 and 6a), which coincided with soil water content reaching lethal levels. In this case the population crashed to below 10 % of its original size, which then had profound effects on the post-drought recovery of the species at the study site. Interestingly, population-level mortality was visibly much lower during 2002 (the second driest year on record with 225 mm), illustrating that small rainfall differences, even among extreme events, can have great ecological significance.

Similar patterns of nonlinearity have been widely observed in response to drought, along with many other climatic cues, and often reflect the presence of discreet thermal or water stress thresholds beyond which normal physiological processes are impaired and cell death occurs. For example, a run of successive 9 days of overnight frosts in July 1984 where screen temperatures dipped to around the freezing point resulted in the widespread mortality of early-successional rainforest species at the Atherton Tablelands in northern Queensland (Fig. 2), which, like most tropical species, are frost sensitive with discrete low temperature thresholds that induce mortality (Prentice et al. 1992). Similar effects were observed in the same area in 2007 (Curran et al. 2010), and it has been suggested that occasional severe frosts are probably an important factor in determining the distribution of rainforest in Australia (Duff and Stocker 1989). Similar impacts on vegetation were observed in south-western Australian mallee-heathland vegetation during a

Fig. 6 The study site near Wyalong, NSW where populations of *Austrostipa aristiglumis* were tracked through the drought years of 2006–2008. (**a**) The study site in September 2007. Dead, gray tussocks of *A. aristiglumis* are in the background. Annuals showing varying levels of water stress are present in hollows in the foreground. (**b**) Topography and vegetation of the site in 2010 following drought-breaking rains, showing barer slopes dominated by *Panicum* spp. and *Leiocarpa panaetioides* in the foreground and wetter terrace habitats dominated by *A. aristiglumis* in the *centre left*. Extensive flats dominated by *A. aristiglumis* are in the background. [Courtesy of Robert C. Godfree]

record heatwave in which temperatures exceeded 45 °C on successive days (Groom et al. 2004; Fig. 2); leaf mortality in Mediterranean species is very sensitive to small (2–3 °C) differences in temperature (Larcher 2000).

Predicting the impact of specific ECEs on plant populations can be extremely challenging when physiological thresholds are involved. In the case of drought, for example, we are unlikely to know the underlying physiological responses of most wild plant populations to water stress, the level of within-population genetic variation present for stress tolerance, the potential for phenotypic plasticity to allow plants to persist through stressful periods, or even the exact nature of soil water conditions through the rooting zone over time. A secondary complication is that other landscape processes, themselves linked to specific climatic thresholds, can also result in feedback scenarios which magnify the impact of ECEs on plant populations. An excellent example of this occurred during the 1930s Great Drought in the mid-west US, when the loss of plant cover led to destabilisation of the soil surface, resulting in extensive wind erosion and soil drift. In some places, drifts covered woody vegetation altogether, or enhanced drying of the root zone by reducing the permeability of the soil surface to precipitation (Albertson and Weaver 1945). Collectively, these lines of evidence suggest that the past behaviour of plant populations to previous climate events may often be a poor or incomplete guide to their behaviour when faced with new extremes.

Species-Level Responses to ECEs, Ecosystem Change, and the Role of Environmental Heterogeneity

Interspecific differences in climate sensitivity and stress tolerance are probably ubiquitous in ecosystems globally, having been observed in drought-affected tropical rainforests and temperate woodlands, in heatwave-affected mallee-heathland, and in alpine grassy heath, subalpine woodlands and tropical rainforests affected by frost (see Fig. 2), to name but a few. These species differences fundamentally affect the way in which plant assemblages respond to severe abiotic stress, and so it appears safe to conclude that all ecosystems affected by ECEs will likely undergo at least some degree of structural and floristic change, at least temporarily.

As discussed above, ECE-induced mortality is often a key driver of vegetation change, and when stress-sensitive species are lost major shifts in plant community composition can occur. Such changes can also happen surprisingly rapidly and at landscape scales. An excellent example of this occurred in the south-western US during the 1950s, when the ecotone between ponderosa pine (*Pinus ponderosa*) forest and piñon-juniper woodland shifted significantly in response to a severe drought (Allen and Breshears 1998). Here, extensive mortality of *P. ponderosa* in response to acute soil water deficiencies resulted in ecotonal movement of 2 km or more within a <5 year period. The fact that little reestablishment of *P. ponderosa* occurred in formerly occupied parts of its range in subsequent favourable years resulted in the persistence of this ecotonal shift for at least 40 years, demonstrating that brief climatic extremes can have long-term ecosystem-wide consequences. Interestingly, the retreat of *P. ponderosa* to higher, wetter locations was also accompanied by increased fragmentation of remaining forest patches.

Typically, however, the impacts of ECEs on population demography and community composition are more complex and occur over longer timeframes. Most plant species that occur in environments characterised by recurring abiotic stress have one or more demographic mechanisms that enable populations to either survive through, or rapidly recover from, extreme events once conditions ameliorate (e.g., the production of a persistent seedbank). Therefore changes in ecosystem structure and composition often reflect the processes of both mortality and post-event recovery. Both were shown to be of great importance in the *Austrostipa aristiglumis*—dominated semi-arid Australian grassland described above (see Godfree et al. 2011; Fig. 2) between 2006 and 2008, a period of record rainfall deficiencies, critically low soil water availability, record warmth, and elevated evaporation (Murphy and Timbal 2008). While these conditions resulted in mortality of virtually all perennial species at the study site, the subsequent recovery of populations during 2007–2009 reveal some very interesting aspects of plant population responses to ECEs that may well be generalisable to other systems.

First, drought-induced mortality differed greatly among the three dominant species at the site (*A. aristiglumis*, another perennial tussock grass *Panicum prolutum*, and the shrub *Leiocarpa panaetioides*; Fig. 6b), resulting in a considerable shift in the relative abundance of all three species over the study period. In general, *L. panaetioides* suffered much lower mortality than *A. aristiglumis* or *P. prolutum* during the drought, but in contrast to the latter species showed little evidence of population recovery after the break of the drought. Most importantly, however, minor topographic variation (<4 m across the study site) greatly influenced the rates of mortality and recovery of all species, with the composition of the different habitats investigated changing over the drought-recovery transition period (described in detail in Godfree 2012). Furthermore, the mix of strategies adopted by plant species to survive the drought period was highly habitat-dependent, and ranged from high adult survival (mesic habitats) to high post-drought recruitment (more xeric habitats). Populations of *A. aristiglumis* and *P. prolutum* virtually disappeared on the driest terrain, also demonstrating the persistent ecological impacts of transient climatic conditions.

Evidence that topographic, edaphic and microclimatic variability plays a critical role in determining the demographic responses of plant species to ECEs has been widely observed in other studies. For example, Groom et al. (2004) noted that species-level differences in the extent of damage caused by heatwave conditions appeared to be linked to the degree of pre-adaptation of plant species in habitats varying in exposure to heat and desiccation, while similar microclimatic impacts associated with the degree of canopy cover influenced the impact of frost on tropical vegetation in northern Australia (Curran et al. 2010). At a larger scale, Albertson and Weaver (1945) noted that woody vegetation mortality varied widely in different topographic positions under drought conditions, and it is typical for the first plant communities to be stressed under increasingly dry conditions to be those growing on elevated terrain or in shallow soils. These case studies also indicate the potential importance that refugial habitats can play in both maintaining viable plant populations during and after an ECE and by acting as a source of propagules following amelioration of climatic conditions.

The Importance of Complex ECE-Population-Environment Interactions

Above, we focused primarily on the direct impact of acute and chronic extreme abiotic stress on plant population dynamics and some of their impacts at local to landscape scales. In perhaps the majority of cases, however, climatic extremes drive demographic and ecosystem change via interactions that involve a broader range of abiotic and biotic processes. These often involve secondary indirect or feedback mechanisms that amplify the initial direct effects of ECEs on plant populations, leading to even greater ecosystem change.

Among such processes, the amplifying effects of disease and insect attack on forest tree populations under stress from climate change, rising atmospheric CO_2 levels, soil nitrification and other drivers of global change have arguably received the greatest attention. Many recent studies suggest that the mortality rate of mature and old trees has risen in ecosystems globally, with insect attack and disease likely playing an increasingly important role in these systems (Harvell et al. 2002; Kurtz et al. 2008; Anderlegg et al. 2013). Such concern is not new, however: the interactive effects of disease and climate extremes, particularly drought, have been widely implicated in the widespread dieback of forest across southern and eastern Australia over the past century. While climate is thought to be only one of many factors which have led to increased impacts of defoliating insects and pathogens on tree health, the link between extreme or unprecedented drought and subsequent insect or pathogen behaviour has been well documented in some situations.

An excellent example was reported by Pook et al. (1966) and Pook and Forrester (1984), who studied the effects of an acute drought between December 1964 and August 1965 in Canberra, ACT, Australia (Fig. 2) on naturally occurring tree populations. During this drought, Canberra received only 64 mm of rain in the first half of 1965 (the driest such period on record), and also experienced well above-normal pan evaporation. The ensuing exhaustion of soil moisture, which extended for over 30 weeks, severely stressed trees in the region, with many experiencing extensive canopy defoliation, wilting, and bark injury associated with desiccation. Bark fissures and cracks provide favourable oviposition sites for longicorn beetles, and subsequent larval feeding resulted in the girdling, and ultimately death, of the most severely drought-affected trees, well after the drought had broken. Interestingly, there is also some evidence that growth and survival of longicorn beetle larvae is higher in water-stressed than well-watered trees (Caldeira et al. 2002). This case study not only demonstrates that lag-effects can be important in drought-affected systems, but also that differences in susceptibility to insect attack can significantly affect the expression of ECEs at both intraspecific and interspecific scales; here, bark damage and beetle infestation were most prevalent in younger age classes of smooth-barked tree species.

Even more complex interactions occur in systems where plant-climate relationships are affected by a high level of ecosystem sensitivity to other biotic and abiotic variables, especially where rapid transitions between alternative (stable) states can arise. Semi-arid rangelands are perhaps the biome most sensitive to such changes, since they are usually episodic or 'event driven' systems in which plant survival and recruitment disproportionately reflect the occurrence of rare climatic events or sequences of events (Walker 1993). Indeed, failure to realise the critical sensitivity of these systems to changes in land management and climate has, in the past, resulted in the tragic collapse, and, in some case permanent damage, of semi-arid ecosystems in many parts of the world.

This lesson was learned the hard way in the semi-arid rangelands of western NSW (along with much of the rest of semi-arid Australia) during the first half of the twentieth century, a period of generally reduced rainfall punctuated by extended droughts of great severity that affected much of the continent. In this region significant changes in rangeland vegetation occurred following settlement in the mid 1800s (for example woody weed invasion and the loss of palatable species), but it was the 1896–1902 "Federation Drought" which first resulted in severe rangeland degradation as a direct consequence of overstocking and massive loss of perennial vegetation. During this drought (Verdon-Kidd and Kiem 2009), much of inland NSW was affected by extensive wind erosion, drifting sand and dust storms. While rangeland conditions improved somewhat after 1903, chronically dry conditions were again experienced repeatedly in the 1910s–1930s, culminating in the successive extreme drought years of 1943–1945 when much of south-western NSW and northern Victoria suffered the lowest 36-month rainfall on record. During this period rangeland collapse was general, dust storms occurred regularly, and wind- and water-driven soil loss occurred on a massive scale (reviewed in McKeon et al. 2004).

Fortunately, the astute observations of landowners and government officials at the time and decades of subsequent research provide a clear picture of the interactions between grazing, plant cover, and climatic conditions that led to the degradation and subsequent (partial) recovery of western NSW rangelands during and after this period. By far the most important factor throughout the period was livestock grazing pressure, which prior to 1880 had been restricted by water availability. However, an increase in watering points, generally favourable (wet) conditions years in the 1870s and early 1890s, and a great plague of rabbits which migrated across inland NSW in the 1870s and 1880s caused grazing pressure to rise to unsustainable levels. During the ensuing 7-year drought perennial groundcover was reduced to low levels by grazing, while trees and shrubs were ringbarked by starving rabbits. Defoliated grasses lacked the photosynthate reserves to re-allocate growth to roots during dry periods, and suffered extensive mortality. Wind erosion then exposed less fertile and permeable subsoils, which in turn reduced water filtration and increased scalding of the soil surface. Extensive mobilisation of sand during windstorms blasted the remaining perennial vegetation even on better managed

Fig. 7 Extensive surface scalding and loss of vegetation in north-western NSW. Extreme drought combined with poor land management practises resulted in rangeland degradation and erosion on a massive scale between the 1890s and 1940s. While many areas have recovered, there remains widespread evidence of these past events. Near Carinda, NSW, in 2006 (Fig. 2). [Courtesy of Robert C. Godfree]

properties. The destruction of the pre-existing soil crust, which was strengthened by lichen and other microflora, then reduced soil microbial activity and nutrient cycling, which limited vegetation recovery. The resulting collapse of vital ecosystem processes, which even today have not fully recovered (Figs. 2 and 7), demonstrates how the impact of extreme climatic conditions can be augmented by complex landscape-level feedback loops.

As a postscript to this case study, it is interesting to reflect on the factors which led to the dramatic improvement of the NSW rangelands beginning in the 1950s. The most important factor was certainly a shift towards generally higher rainfall, and in particular the exceptionally wet years of 1950, 1955–1956, and 1973–1976, when annual precipitation was at or above 90th percentile totals over most of eastern Australia. A decline in rabbit density following the release of the myxoma virus in 1950, along with the new ability to move stock during drought periods using road trains allowed perennial vegetation (including woody weeds, which themselves are considered a driver of rangeland degradation) to re-establish across most of the region. Soil stabilisation was also assisted by coordinated engineering and land management strategies aimed at reducing water movement and re-establishing vegetation, and probably by a general decline in wind speeds. Severe droughts still result in degradation, but there is hope that the unique combination of climatic and anthropogenic processes that led to the events of the 1890s–1940s may not again be repeated.

Extreme Events in Aquatic Systems

Above, we have focused on the impacts of extreme climatic events on terrestrial systems. However, extreme events, such as floods and droughts, can also be important drivers of the structure, composition, assembly, and function of aquatic systems (Lytle and Poff 2004; Naiman et al. 2008). Ecologically, aquatic ecosystems are extremely diverse, and unsurprisingly, the impacts and indeed the very nature of ECEs vary widely among these different systems; here we focus on the impacts of extreme hydroclimatic events on freshwater aquatic ecosystems.

Extreme hydrological events are a natural part of many aquatic ecosystems, some of which may even rely on both extreme floods and droughts to maintain their composition and structure over time (Lytle and Poff 2004). The study of extreme events in aquatic systems is complicated by the tendency for associated hydrological regimes to reflect not only prevailing climatic or meteorological conditions at a given location but also a myriad of water balance-related processes operating at site to landscape scales. For example, river flow rates, groundwater availability, and the timing of seasonal inundation often substantially lag changes in rainfall, temperature, evapotranspiration and land surface runoff rates. Some aquatic ecosystems also require a very long historical record to characterise long-term variability in flow, particularly in highly variable or intermittent systems (Lake 2011; Reid and Ogden 2006), a difficulty compounded by anthropogenic river regulation, which alters the statistical distribution and timing of flows, directly uncouples meteorological and hydrological drought, and generates new or "non-natural" types of extreme events. In many cases, the frequency and severity of extreme events have actually been *reduced* by river regulation and water extraction for human purposes (Walker et al. 1995; McMahon and Finlayson 2003; Lytle and Poff 2004; Vivian et al. 2014a).

Numerous indices have been developed to quantify both the hydrological regime of an aquatic ecosystem and the thresholds that define extreme events; events can be extreme in terms of magnitude, frequency, seasonality, duration, or spatial factors. As with climatological parameters there will always be a selection of events that can be classified as statistically extreme (e.g., beyond the 95 % percentile), but we are usually interested in those that elicit a biotic response in which the adaptive capacities of an organism are exceeded. This distinction is especially relevant for aquatic ecosystems, in which many species have evolved traits (e.g., life history, morphological plasticity, different regeneration or reproductive strategies) that allow them to survive extreme disturbances (Lytle and Poff 2004). Particular adaptations can carry costs as well as benefits, and changing the timing, magnitude, duration and predictability of extreme events can shift the balance to cause the costs of the adaptation to outweigh the benefits (Lytle and Poff 2004), thus inevitably affecting species survival and persistence.

Case Study: The Murray-Darling Basin and the Millennium Drought

Australia's Murray-Darling Basin (MDB) is a region where the impacts of extreme events on aquatic vegetation have been highly visible in recent years. While inflows in the MDB have traditionally exhibited high variability on annual to multi-decadal timeframes, between 1997 and 2009 (the "Millennium Drought") much of south-eastern Australia and the MDB suffered from severe meteorological drought, with many areas recording record low rainfall in this period. Low rainfall was exacerbated by high evaporation rates (due to above average temperatures), which resulted in an even larger decline in runoff and the establishment of extreme (90th–95th percentile) or even unprecedented hydrological drought across much of the region (Leblanc et al. 2012; Cai and Cowan 2008). The Millennium Drought was finally broken by flooding rains during the La Nina years of 2010–2011 (Leblanc et al. 2012).

The nature of these extremes can be illustrated by stream flow records from Yarrawonga weir, located approximately 1992 km upstream from the mouth of the Murray River, and the impacts on downstream vegetation in the Barmah-Millewa Forest (BMF) (Fig. 2). The BMF contains a range of high conservation-value flood-plain vegetation, including extensive stands of the iconic *Eucalyptus camaldulensis* (river red gum), many of which are flooded when flow in the Murray River exceeds the channel capacity of 10,600 mL day^{-1} at Yarrawonga weir. Prior to the regulation of the Murray River in the mid-1930s, in most years BMF experienced deep winter-spring floods and a generally dry summer-autumn period. Flow during the latter part of the Millennium Drought was very low, remaining approximately at or below channel capacity (<ca. 11,000 mL day^{-1}) for a continuous period of over 1600 days between February 2006 and 2010, resulting in the absence of significant flooding in the BMF. This was by far and away the longest duration of low flow on record, the two previous being 626 days (December 1943 to August 1945) and 582 days (March 1940 to October 1941).

Interestingly, the statistical 'extremeness' of daily flows during the Millennium Drought and the subsequent La Nina flood peaks is strongly influenced by the choice of reference period. For example, when compared to the post-regulation reference period of 1960–1990, daily flow during the last 2 years of the Millennium Drought (2008–2010) only occasionally dropped below the fifth percentile (Fig. 8a). In contrast, when compared to the pre-regulation period of 1905–1935, daily flow was well below the fifth percentile for multiple lengthy periods, including during every winter and spring (i.e., June to November) from 2006 to 2009 (Fig. 8b). Collectively, daily winter and spring flows during the Millennium Drought were thus more extreme when compared with the pre-regulation period than the post-regulation period.

Multiple floods occurred from late 2010, ending the long-term drought conditions (Fig. 8). The flood peaks that occurred in winter-spring 2011 and 2012 were well within the normal bounds of daily low rates, but other peaks were clearly unusual.

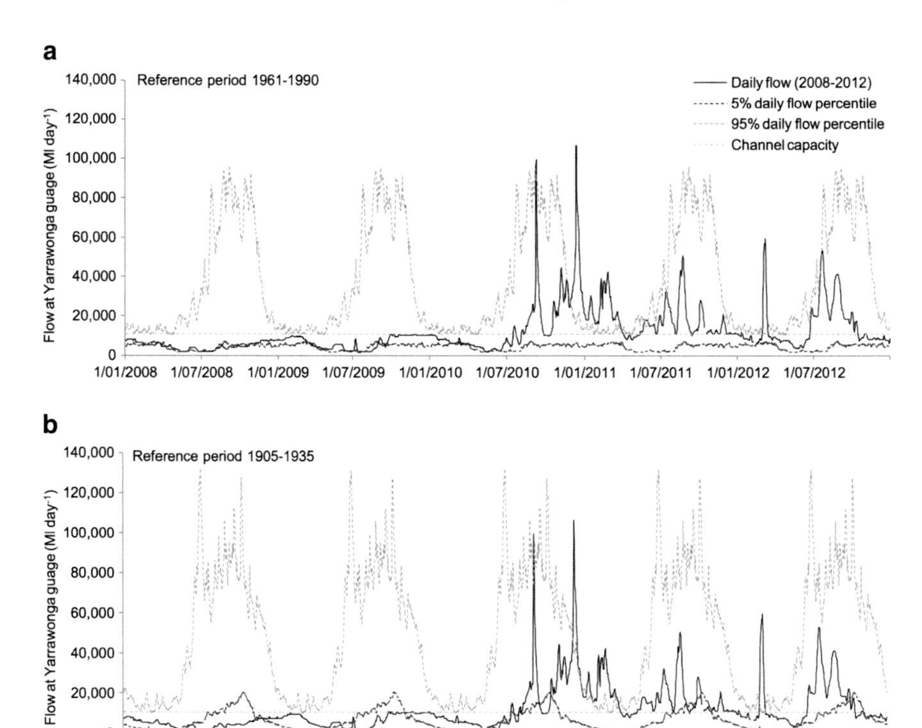

Fig. 8 Daily flow at Yarrawonga weir (*black line*) during the final 2 years of the Millennium Drought (2008–2010) and the floods of 2010–2012. *Dotted grey line* indicates the bank threshold of ca. 10,600 ml/day above which flooding occurs in the downstream Barmah-Millewa Forest. Flow is shown compared to 95 % (*blue line*) and 5 % (*red line*) flow percentiles of two reference periods: (**a**) 1960–1990 (post-regulation) and (**b**) 1905–1935 (pre-regulation)

In particular, the unseasonal flooding that occurred in summer 2010–2011 can be considered extreme with respect to both reference periods, extending for several months above the 95 % percentile of daily flows. In particular, peak flows in March of both 2011 and 2012 were the two highest March flood peaks on record, and occurred at a time when river flow is normally below channel capacity and floodplains are consequently dry.

Given the extremes in flow seasonality, depth and duration described above, it is not surprising that these drought and flood events had a significant impact on major species and ecosystems across the BMF. Perhaps most dramatic was the widespread dieback of *E. camaldulensis*, a situation also observed widely across the Murray-Darling Basin (Fig. 9). During latter stages of the drought *E. camaldulensis* located away from the Murray River were in poorest condition (Cunningham et al. 2009), reflecting the decline in flood extent within the forest. Interestingly, Horner et al. (2009) showed that annual *E. camaldulensis* mortality increased substantially between 1996 and 2007, but that the impact was limited to higher density stands.

Fig. 9 Extreme drought stress in *Eucalyptus camaldulensis* (river red gum) during the Millennium Drought. Dieback was observed in river red gum populations across much of south-eastern Australia during the drought as a result of extremely low rainfall combined with water extraction for human use. [Courtesy of Robert C. Godfree]

They also noted that the impact of the Millennium Drought exceeded that of drought periods in the past, probably due to record high temperatures, low rainfall, a reduction in water-table depth, and the increasing removal of water from the Murray River for irrigation (Horner et al. 2009).

The impact of these extreme flow regimes on *Juncus ingens* (giant rush; Fig. 10a) in the BMF also demonstrates the role that extreme events can play in maintaining vegetation composition, and the changes that can occur following river regulation. *J. ingens* is a very tall (2–3 m) native rush which has expanded to form extensive near-monospecific stands across much of the BMF, where it is considered a key cause of degradation of treeless floodplains dominated by *Pseudoraphis spinescens* (Moira grass). *J. ingens* is intolerant of sustained submergence during deep winter/spring floods, and so has increased dramatically in distribution and abundance following river regulation (Vivian et al. 2014b).

During the latter stages of the Millennium Drought, soil water deficits were so severe on some floodplains that death of mature *J. ingens* was observed (Mayence et al. 2010; Fig. 10a) and at least some invasive populations suffered significant declines in stem density. Even more dramatic impacts were observed during and after the 2010–2012 floods, when extensive stem death of *J. ingens* was recorded in lakes and open plains across much of the forest. Stem mortality was greater, and post-flood recovery lower, in areas that experienced deeper floods, such as along the edge of Barmah Lake (Fig. 10a): as shown in Fig. 10b, recovery of *J. ingens* was negligible in areas submerged by more than 2.5 m of water.

a

b

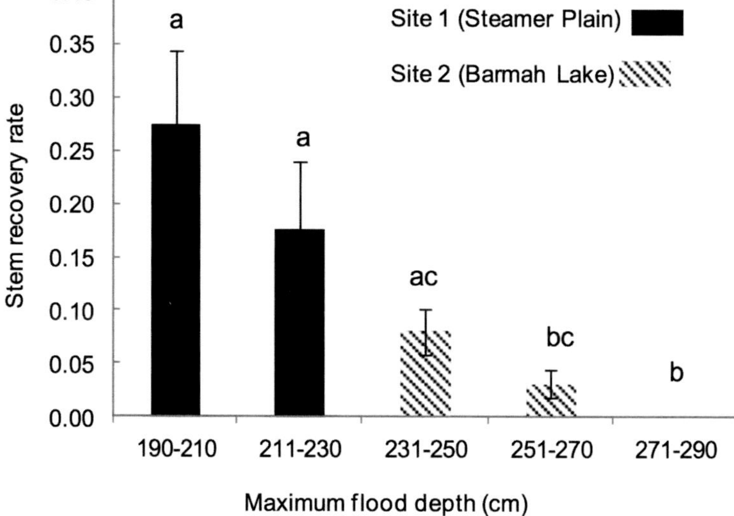

Fig. 10 Impact of flood depth on *Juncus ingens* survival and recovery at Barmah-Millewa Forest during the 2010–2012 floods. Deeper floods result in stems more likely being submerged for longer periods. (**a**) Stem die back (*red arrow*) following prolonged submergence. This site is along the edge of Barmah Lake where *J. ingens* has encroached over several decades to become invasive. Taller species such as *Phragmites australis* (*green arrow*) have remained emergent above flood water depths and have survived the extreme flooding. (**b**) Comparison of stem recovery rates ((new stems + surviving stems)/old stems) in response to different flood depths. *Bars* with *different letters* are significantly different (ANOVA). [Courtesy of Lyndsey M. Vivian]

The results of this study demonstrate that extreme events can play a critical role in determining the structure and composition of aquatic vegetation, and may even be essential for maintaining ecosystem health. The impact of flooding and drought on *J. ingens* populations strongly suggest that periods of extremes in river flow, especially deep floods, historically limited the invasive potential of this species. These constraints were reduced by river regulation, with events of sufficient extremity to halt the invasion of *J. ingens* now occurring only occasionally.

ECES and the Response of Plant Populations to Climate Change

The case studies above demonstrate that climate extremes have the potential to profoundly affect wild plant populations in both terrestrial and aquatic environments. Demographic changes induced by abiotic stress often occur in a rapid, non-linear manner, and involve persistent shifts in vegetation structure and composition that manifest at plant community to ecosystem scales. However, while the case studies above primarily involve the impact of short-term extreme events which are followed by a return to prior, more favourable conditions, recent developments in our understanding of the nature of anthropogenic climate change has seen an increased focus among ecologists and evolutionary biologists on the role of ECEs in changing climates.

Most plant species have a geographic range that is delineated at least in part by their physiological ability to tolerate climatic stress, known as their climatic envelope. As discussed previously, when climatic extremes move plant populations outside of this envelope, fitness declines and populations may contract. However, because climate varies across a species' range, populations from different parts of that range experience different levels of abiotic stress which may or may not lead to demographic change. Consider, for example, a plant species that occurs within a climatic envelope delineated by upper and lower thresholds for a specific climatic variable, such as rainfall, as shown in the diagram in Fig. 11a. In this very simple, stylised model, populations persist within the geographic region that experiences climatic conditions that lie within these thresholds (populations A–C, Fig. 11a) for a proportion of the time sufficient to maintain stable or positive population growth.

Fig. 11 (continued) C becomes range-core, and a new population could become established at D via migration. However, population B faces extinction if it cannot adapt. (**c**) Climatic conditions experience by a range-core population with stable long-term climate. Even extreme events (marked by the *arrow*) may not fall outside the physiological limits of the population. (**d**) Climatic conditions experienced by a range-edge population, with extreme events (*arrows*) potentially exceeding the population's physiological limits. (**e**) Potential scenario for a range core population able to persist even in the presence of climate change. (**f**) Potential scenario for a range edge population which faces an increasing number of damaging climatic events and possible extinction

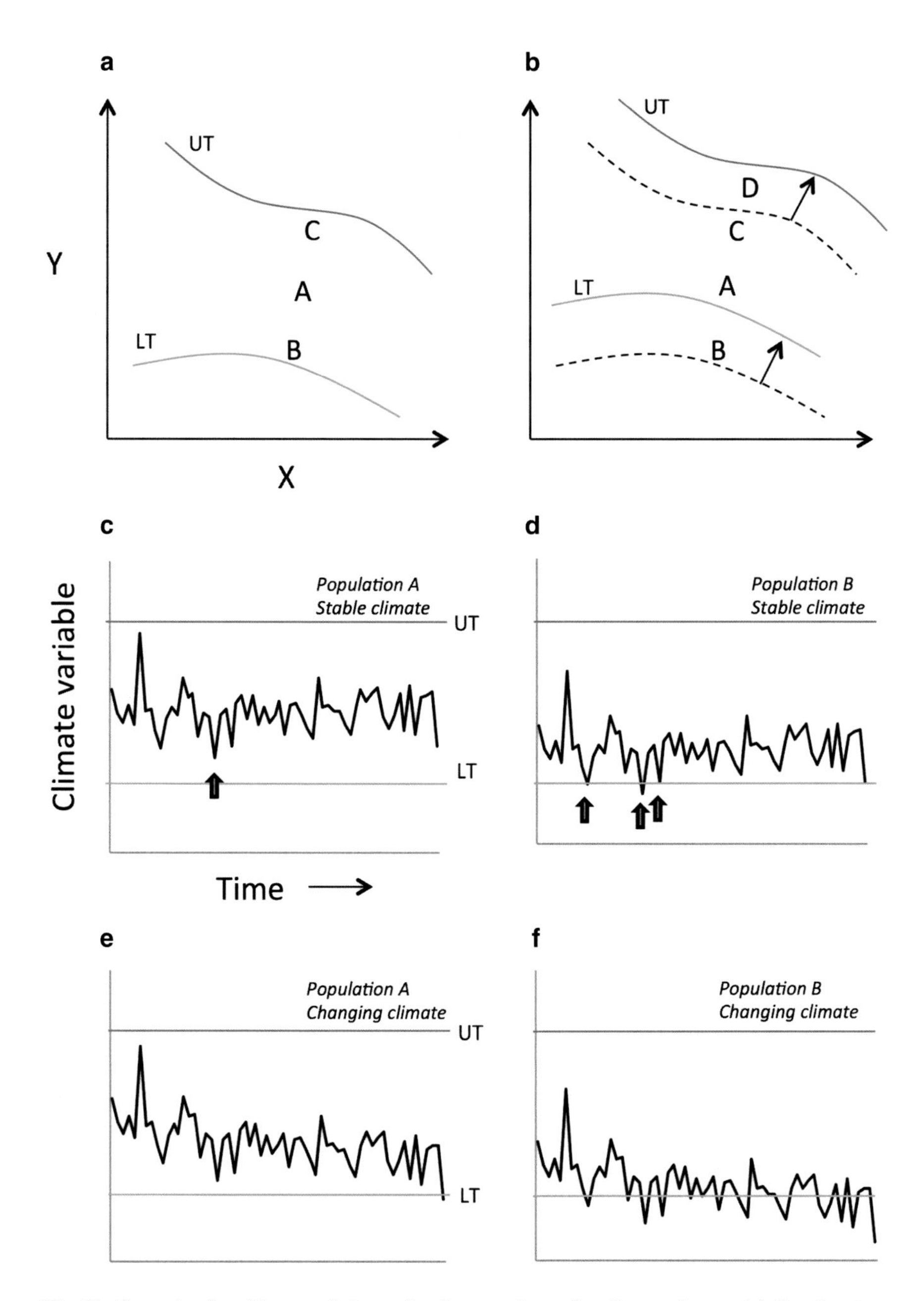

Fig. 11 Scenarios faced by populations of a plant species under climate change. (**a**) Simple representation of a situation where the range of a species is determined by a single climate variable. The range is shown in two spatial dimensions (X and Y) and is bounded by the species' upper and lower climate thresholds (UT and LT). Population A exists in the core of the species range, while B and C exist towards the range edge. (**b**) Scenario in which the species' climatic envelope moves in space, shown by the *arrows*. Population A now lies near the edge of the climatic envelope,

Towards the centre of their species range, populations (A in Fig. 11a) generally experience favourable climatic conditions which rarely, if ever, exceed the stress tolerance characteristic of the population (Fig. 11c). Extreme events, while a statistical phenomenon (Fig. 11c), are less likely to be of sufficient severity to affect the demography of resident populations, especially in a catastrophic manner. In contrast, environmental conditions in range - edge areas (B and C in Fig. 11a) are more marginal, and populations tend to be subjected to a higher frequency of events that are near to or exceed their physiological stress thresholds (Fig. 11d). Although populations can to a certain degree adapt to these conditions by means of life history avoidance strategies (e.g., the production of a persistent seedbank) and the ability to alter phenotype in response to environmental cues (phenotypic plasticity), at a certain point adaptive limits are reached, populations fail to maintain positive growth rates, and ultimately go extinct.

There are good reasons to think that a high frequency of ECEs and associated episodes of extensive plant mortality (c.f., Tigerstedt 1994) play a particularly important role in the evolutionary dynamics and viability of range-edge plant populations. As discussed above, many of the best examples of rapid vegetation mortality and change associated with ECEs do come from range–edge environments (e.g., Albertson and Weaver 1945; Allen and Breshears 1998), although it should be noted that there are certainly exceptions (e.g., Godfree et al. 2011). Due to their typically small size, fragmented distribution and low fecundity (Bridle and Vines 2007), range-edge populations are more likely to suffer adverse evolutionary consequences of ECE-induced bottlenecks, and in particular the deleterious effects of inbreeding depression, genetic drift, and Allee effects (Bridle and Vines 2007). Indeed, these processes, and the associated dynamics of migration-selection balance in which adaptive potential is maintained without swamping of advantageous alleles (Phillips 1996; Bridle and Vines 2007), are often invoked to explain the presence of relatively stable range limits that are observed in many plant species globally (Hoffmann and Hercus 2000).

In a changing climate, many plant populations face the likelihood of being exposed to increasing abiotic stress as their climate envelope undergoes geographic movement (Fig. 11b). For some populations (e.g., population C in Fig. 11a, b), climatic conditions may become more favourable, even leading to the establishment of new founder populations in areas previously located outside the species range (e.g., population D in Fig. 11b). Others, especially those with broad climatic envelopes which encompass past and future climate regimes, may be relatively unaffected by the changing climate, at least temporarily (e.g., Fig. 11e). These populations may adapt via expression of advantageous plastic phenotypes, although the behaviour of plasticity traits in truly novel climates may be disrupted by climatic extremes and remains poorly understood (Chevin et al. 2010). So-called rear or trailing edge populations (B in Fig. 11a, b) face exposure to climatic conditions that lie outside their physiological limits (Fig. 11f), and for these populations evolutionary adaptation becomes necessary if extinction is to be avoided (Kuparinen et al. 2010).

The extent to which evolutionary processes are likely to allow rear-edge plant populations to track climatic change remains a contentious issue (Kuparinen et al.

2010). On the one hand, there is abundant evidence that ongoing physical stress can shift the means of adaptive traits by means of directional selection (Reznick and Ghalambor 2001), even if the intensity of selection is not high (Hoffmann and Hercus 2000). On the other hand, clear evidence for a dominant role of evolution in species responses to past episodes of climate change remains lacking (Gienapp et al. 2008); most species appear to track their climatic envelopes closely during periods of climatic change. Evidence from invasion biology also suggests that invasions rarely occur in the absence of pre-adaptation of genotypes to newly encountered climatic regimes (Bridle and Vines 2007). Nevertheless, there are good reasons to once again expect that ECEs will play a significant role in the evolution of plant genomes under changing climatic conditions.

First, there is evidence that populations subjected to periodic disturbance and episodes of high mortality, such as occur during ECEs, might adapt most rapidly to climate change, since low mortality in established cohorts can slow the production and establishment of better adapted juvenile genotypes. This process has been modelled for populations of the tree species *Pinus sylvestris* and *Betula pendula*, where an adaptive lag in response to climate change was reduced by high adult mortality (Kuparinen et al. 2010). Second, stressful events can trigger the expression of new phenotypes within a population, which may then increase in frequency and maintain expression even under ameliorated growing conditions (Hoffmann and Hercus 2000). Such events may even increase mutation and recombination rates and hence the production of entirely new genotypes, and may also affect the heritability of traits under selection (Hoffmann and Merilä 1999). Finally, since population differentiation is also facilitated by low rates of outcrossing, ECEs may play an important role in local adaptation by reducing fecundity and hence the introgression of maladaptive alleles into the genomes of range-edge plant populations.

However, periods of elevated physiological stress that lead to high plant mortality may also hinder the optimal tracking of phenotypes to prevailing climatic conditions. First, small population sizes reduce the likelihood of *in situ* generation of better-adapted genotypes (Sgrò et al. 2011), and in general evolutionary change seems to primarily reflect the selective pressures that occur in environments that are experienced by the majority of individuals in a population (Bridle and Vines 2007). Second, extreme events may cause selective sweeps that reduce the diversity of genetic traits that mediate the organismal response to changes in climate, resulting in elevated rates of inbreeding and reduced fitness. Such effects are likely to be most severe in rear-edge populations, where extreme events are likely to have the biggest demographic impact on populations. Third, extreme events may drive rapid selection for traits that are adaptive under stressful conditions but maladaptive once favourable growing conditions return and other drivers such as disease and competition become more important. Finally, the evolvability of traits under climatic change is contingent on the presence of heritable genetic variation, and past extreme events may have already driven advantageous alleles to fixation in range-edge populations, thus reducing the capacity for further adaptive change (Davis et al. 2005).

As discussed above, a critical feature of ECEs is that they tend to affect populations in a nonlinear manner, with mortality rates increasing rapidly once certain physiological thresholds have been exceeded. This may have important implications for plant evolution under climate change, because it suggests that topoedaphic factors that alter the immediate abiotic environment experienced by individual plants can have an immense impact on plant survival during extreme events. Indeed, in a heterogeneous environment we might expect ECEs to select plants that survive not by virtue of being better adapted to climatic extremes, but simply because they grow in a marginally more favourable environment. This was observed in the *Austrostipa aristiglumis* populations studied by Godfree et al. (2011), where minor differences in soil water availability was the dominant predictor of plant survival during a period of extreme drought, and where no increase in drought tolerance was observed among drought survivors. Under such circumstances, which are probably common, ECEs may act more as an agent of random rather than selective mortality.

A Role for Genomics in the Study of the Past and Future Behaviour of Climatic Extremes

Genomics can inform how ECEs have effected past demographic and evolutionary history of plant populations. Indeed, both selection and demography history should be considered jointly when inferring the history of a population based on observed genetic variation (Siol et al. 2010). Past demographic and selection events can result in similar signals and, until recently, it has been difficult to tease apart competing hypotheses. The primary difference is that demographic events should affect neutral variation across the genomic in a similar manner, while selective forces will only affect genes under selection. Genomics provides an opportunity to tease apart competing hypotheses by providing large datasets in which to explore patterns of variation across the genome. Specifically, by sampling a large number of unlinked genes from across genome, it is possible to identify patterns that likely result from selection pressures versus demographic events (Akey et al. 2004).

Demographic History

Genomic data can be DNA sequences or, commonly, single nucleotide polymorphisms (SNPs). Genomics allows the identification of a large number (100s–1000s) of SNPs from across the genome. SNPs allow more precise estimates of population history, compared to microsatellite markers, due to more predictable mutation rates (Brumfield et al. 2003). SNPs however can suffer from ascertainment bias that can be avoided by using DNA sequence data directly.

Large amounts of genome-wide data allows for estimating several parameters associated with more complex models of population history, including migration rates, timing and strength of bottlenecks, and expansions (Stoneking and Krause 2011).

The application of coalescent theory, or patterns of common ancestry (Kuhner 2009), to genomic datasets allows a more accurate picture of population history as there is typically variation among gene trees and the more unlinked genes examined allows for a more complete picture of population history.

The application of coalescent theory to many nuclear genes has been limited by computational challenges. Recent methods, such as pairwise sequentially Markovian coalescent (PSMC) modelling have shed light on complex human demographic history (Li and Durbin 2011). Similar methods applied to polar bears have informed past fluctuations in effective population size (N_E) in relation to past climatic events (Miller et al. 2012). More commonly, Bayesian and Approximate Bayesian (ABC) approaches are used to fit demographic models to genomic data. For example, Ross-Ilbarra et al. (2008) reveal past fluctuations in effective population size in populations of *Arabidopsis lyrata* in relation to past glacial cycles.

Evolutionary History

Importantly, fitting demographic models to genomic data provide the necessary backdrop to explore signals of adaptation (Ross-Ilbarra et al. 2008; Siol et al. 2010). Siol et al. (2010) provide a summary of plant studies that have used a coalescent approach to fitting demographic models to study selection.

Genomics can provide insights into past extreme events that have resulted in directional selection on a gene with a relatively large effect. A primary approach for identifying genes under selection is 'outlier' tests which identify loci that are out of the range of expected divergence (often F_{ST} based). Expected divergence of neutral variation is heavily dependent on demographic history and therefore the use of demographic models in conjunction with outlier test is an important advance in understanding both demographic and evolutionary history of a population. Genome wide association studies (GWAS) take this a step further and look for correlations among 'outlier' loci purportedly under selection and certain traits such as drought tolerance. GWAS have had a major impact on human medical research and other research areas, including plant studies, have been quick to follow. However, it is important to note that some genes may be adaptive under certain environmental conditions and neutral under other conditions (genotype × environment interactions). Therefore, signals of past adaptive change due to one ECE may not convey adaptive traits to a future ECE. Additionally, phenotypes are often the product of many genes of small effects and signals of selection on these traits can be much harder to detect.

Phenotypic plasticity is yet another way in which plants can temporarily respond to ECEs. While phenotypic plasticity can be heritable, little is still known about the genetic basis (Anderson et al. 2011). Transcriptomics is allowing us to understand more about gene expression, and the effect of differing gene expression on phenoytpes. In order to inform ECEs, transcriptomics requires the sequencing of RNA during the extreme event, ideally from the same individual, to provide insight into how expression may help plants survive. This may be feasible to simulate in a laboratory setting/common garden experiment once regions of the genome hypothesized to be responsible for potential adaptive phenotypes have been identified.

Future Directions and Challenges

The study of climatic extremes has yielded profound insights into the way in which plant populations survive in unpredictable and hostile environments. In many ways, however, this field of research is still in its infancy, and many of the potential impacts of ECEs on the demography and evolution of plant populations remain poorly understood. Some key questions that have yet to be resolved include:

1. *Which plant species, communities and ecosystems are most resilient to demographic change during climatic extremes*? Although an appraisal of the available literature suggests that certain biomes are particularly susceptible to ECEs, or even require them to maintain structure or composition, such as some aquatic ecosystems, few comparative studies have addressed this fundamental question. This is largely due to the difficulties in conducting experiments that simulate ECEs, and so further development of cost-effective techniques for performing such experiments in natural plant communities is urgently required.
2. *What are the ecological and evolutionary dynamics of plant populations following ECEs and when does recovery fail*? The majority of studies that have comprehensively documented the impact of climatic extremes on the demography of wild plant populations have focused on their immediate impact on plant performance and mortality. There is a need to better understand the factors that limit the fitness of population under more favourable conditions that follow ECEs, such as disease, competition and predation. These factors may be at least as important as the direct impacts of the original climatic event itself.
3. *What is the role of ECEs as drivers of evolutionary change?* The evolutionary consequences of ECEs remain virtually unstudied for most groups of organisms. In particular, little is known about the amount and heritability of genetic diversity that exists for climate-sensitive traits in range-core and range-edge plant populations, nor how these are affected by episodes of climate-induced mortality. Lessons can probably best be learnt from studying organisms present in ecosystems which are already exposed to frequent ECEs.
4. *How will the nature of complex ECEs evolve under anthropogenic climate change, and what will the impacts be on plant communities*? Resolving this question is one of the most significant ecological challenges of our time. While many climatic extremes are expected to increase in frequency and severity under climate change, their impact on plant communities often depend not only on climate but on other phytosociological, biophysical and anthropogenic factors. As outlined above, the degree to which drought affects plant populations can be affected by absolute rainfall deficiencies, atmospheric temperatures,

plant population structure and density, soil composition, topography, the activities of grazing animals, disease, and atmospheric CO_2 concentration. Resolving questions of this kind will require an increasingly integrative approach involving a range of disciplines associated with the broader study of global change in general.

References

Akey JM, Eberle MA, Rieder MJ, Carlson CS, Shriver MD, Nickerson DA, Kruglyak L (2004) Population history and natural selection shape patterns of genetic variation in 132 genes. PLoS Biol 2, e286

Albertson FW, Weaver JE (1944) Nature and degree of recovery of grassland from the great drought of 1933 to 1940. Ecol Monograph 14:393–479

Albertson FW, Weaver JE (1945) Injury and death or recovery of trees in prairie climate. Ecol Monograph 15:393–433

Allen CD, Breshears DD (1998) Drought-induced shift of a forest-woodland ecotone: rapid landscape response to climate variation. Proc Natl Acad Sci U S A 95:14839–14842

Allen CD, Macalady AK, Chencouni H, Bachelet D, McDowell N, Vennetier M, Kitzberger T, Rigling A, Breshears DD, Hogg EH (Ted), Gonzalez P, Fensham R, Zhang Z, Castro J, Demidova N, Lim J-H, Allard G, Running SW, Semerci A, Cobb N (2010) A global review of drought and heat-induced tree mortality reveals emerging climate risks for forests. Forest Ecol Manag 259:660–684

Anderlegg WRL, Kane JM, Anderlegg LDL (2013) Consequences of widespread tree mortality triggered by drought and temperature stress. Nat Clim Change 3:30–36

Anderson JT, Willis JH, Mitchell-Olds T (2011) Evolutionary genetics of plant adaptation. Trends Genet 27:258–266

Benson LV, Berry MS, Jolie EA, Spangler JD, Stahle DW, Hattori EM (2007) Possible impacts of early-11th-, middle-12th-, and late-13th-century droughts on western Native Americans and the Mississippian Cahokians. Q Sci Rev 26:336–350

Bridle JR, Vines TH (2007) Limits to evolution at range margins: when and why does adaptation fail? Trends Ecol Evol 22:140–147

Brouwers NC, Mercer J, Lyons T, Poot P, Veneklaas E, Hardy G (2012) Climate and landscape drivers of tree decline in a Mediterranean ecoregion. Ecol Evol 3:67–79

Brumfield RT, Beerli P, Nickerson DA, Edwards SV (2003) The utility of single nucleotide polymorphisms in inferences of population history. Trends Ecol Evol 18:249–256

Cai W, Cowan T (2008) Evidence of impacts from rising temperature on inflows to the Murray-Darling Basin. Geophys Res Lett 35:1–5

Caldeira MC, Fernandéz JT, Pereira JS (2002) Positive effect of drought on longicorn borer larval survival and growth on eucalyptus trunks. Ann For Sci 59:99–106

Chen I-C, Hill JK, Ohlemüller R, Roy DB, Thomas CD (2011) Rapid range shifts of species associated with high levels of climate warming. Science 333:1024–1026

Chevin L-M, Lande R, Mace GM (2010) Adaptation, plasticity, and extinction in a changing environment: towards a predictive theory. PLoS Biol 8, e1000357

Ciais P, Reichstein M, Viovy N, Granier A, Ogée J, Allard V, Aubinet M, Buchmann N, Chr B, Carrara A, Chevakkier F, De Noblet N, Friend AD, Friedlingstein P, Grünwald T, Heinesch B, Keronen P, Knohl A, Kinner G, Loustau D, Manca G, Matteucci G, Miglietta F, Ourvical JM, Papale D, Pilegaard K, Rambal S, Seufert G, Soussana JF, Sanz MJ, Schulze ED, Vesala T, Valentini R (2005) Europe-wide reduction in primary productivity caused by the heat and drought in 2003. Nature 427:529–533

Condit R, Hubbell SP, Foster RB (1995) Mortality rates of 205 neotropical tree and shrub species and the impact of a severe drought. Ecol Monograph 65:419–439

Cramer W, Bondeau A, Woodward FI, Prentice IC, Betts RA, Brovkin V, Cox PM, Fisher V, Foley JA, Friend AD, Kucharik C, Lomas MR, Ramankutty N, Sitch S, Smith B, White A, Young-Molling C (2001) Global response of terrestrial ecosystem structure and function to CO2 and climate change: results from six dynamic global vegetation models. Glob Chang Biol 7:357–373

Cunningham S, Mac Nally R, Read J, Baker P, White M, Thomson J, Griffioen P (2009) A robust technique for mapping vegetation condition across a major river system. Ecosystems 12: 207–219

Curran TJ, Reid EM, Skorik C (2010) Effects of a severe frost on riparian rainforest restoration in the Australian wet tropics: foliage retention by species and the role of forest shelter. Restoration Ecol 18:408–413

Davis MB, Shaw RG, Etterson JR (2005) Evolutionary responses to changing climate. Ecology 86:1704–1714

Dieleman WIJ, Vicca S, Dijkstra FA, Hagedorn F, Hovenden MJ, Larsen KS, Morgan JA, Volder A, Beier C, Dukes JS, King J, Leuzinger S, Linder S, Luo Y, Oren R, de Angelis P, Tingey D, Hoosbeek MR, Janssens IA (2012) Simple additive effects are rare: a quantitative review of plant biomass and soil process responses to combined manipulations of CO_2 and temperature. Glob Chang Biol 18:2681–2693

Doblas-Miranda E, Martínez-Vilalta J, Lloret F, Álvarez A, Ávila A, Bonet FJ, Brotons L, Castro J, Yuste JC, Díaz M, Ferrandis P, García-Hurtado E, Iriondo JM, Keenan TF, Latron J, Llusià J, Loepfe L, Mayol M, Moré G, Moya D, Peñuelas J, Pons X, Poyatos R, Sardans J, Sus O, Vallejo VR, Vayreda J, Retana J (2014) Reassessing global change research priorities in Mediterranean terrestrial ecosystems: how far have we come and where do we go from here? Glob Ecol Biogeogr 24:25–43

Duff GA, Stocker GC (1989) The effects of frosts on rainforest/open forest ecotones in the highlands of north Queensland. Proc R Soc QLD 100:49–54

Gerten D (2013) A vital link: water and vegetation in the Anthropocene. Hydrol Earth Syst Sci 17:3841–3852

Gienapp P, Teplitsky C, Alho S, Mills A, Merilä J (2008) Climatic change and evolution: disentangling environmental and genetic responses. Mol Ecol 17:167–178

Godfree RC (2012) Extreme climatic events as drivers of ecosystem change. In: Mahamane A (ed) Diversity of ecosystems. InTech Publishers, Rijeka, pp 339–366

Godfree RC, Lepschi B, Reside A, Bolger T, Robertson B, Marshall D, Carnegie M (2011) Multiscale topoedaphic heterogeneity increases resilience and resistance of a dominant grassland species to extreme drought and climate change. Glob Chang Biol 17:943–958

Godfree RC, Robertson BC, Gapare WJ, Ivković M, Marshall DJ, Lepschi BJ, Zwart AB (2013) Nonindigenous plant advantage in native and exotic Australian grasses under experimental drought, warming, and atmospheric CO_2 enrichment. Biology 2:481–513

Groom PK, Lamont BB, Leighton S, Leighton P, Burrows C (2004) Heat damage in sclerophylls is influences by their leaf properties and plant environment. Ecoscience 11:94–101

Hannah L, Midgley GF, Lovejoy T, Bond WJ, Bush M, Lovett JC, Scott D, Woodward FI (2002) Conservation of biodiversity in a changing climate. Conserv Biol 16:264–268

Harvell CD, Mitchell CE, Ward JR, Altizer S, Dobson AP, Osfeld RS, Samuel MD (2002) Climate warming and disease risks for terrestrial and marine biota. Science 296:2158–2162

Heberger M (2012) Australia's millennium drought: impacts and responses. In: Gleick PH (ed) The world's water volume 7: the biennial report on freshwater resources. The World's Water, pp. 97–125, doi:10.5822/978-1-61091-048-4_5

Hodell DA, Brenner M, Curtis JH (2005) Terminal Classic drought in the northern Maya lowlands inferred from multiple sediment cores in Lake Chichancanab (Mexico). Q Sci Rev 24: 1413–1427

Hoffmann AA, Hercus MJ (2000) Environmental stress as an evolutionary force. Bioscience 50:217–226

Hoffmann AA, Merilä J (1999) Heritable variation and evolution under favourable and unfavourable conditions. Trends Ecol Evol 14:96–101
Hope PK (2006) Projected future changes in synoptic systems influencing southwest Western Australia. Climate Dynam 26:765–780
Horner GJ, Baker PJ, Mac Nally R, Cunningham SC, Thomson JR, Hamilton F (2009) Mortality of developing floodplain forests subjected to a drying climate and water extraction. Glob Chang Biol 15:2176–2186
Hovenden MJ, Wills KE, Vander Schoor JK, Williams AL, Newton PCD (2008) Flowering phenology in a species-rich temperate grassland is sensitive to warming but not elevated CO_2. New Phytol 178:815–822
IPCC (2012) Managing the risks of extreme events an disasters to advance climate change adaptation. A special report of working groups I and II of the Intergovernmental Panel on Climate Change [Field CB, Barros V, Stocker TF, Qin D, Dokken DJ, Ebi KL, Mastrandrea MD, Mach MJ, Plattner GK, Allen SK, Tignor M, Midgley PM (eds) Cambridge University Press, Cambridge, p. 582]
Katz RW, Brown BG (1992) Extreme events in a changing climate: variability is more important than averages. Clim Change 21:289–302
Katz RW, Brush GS, Parlange MB (2005) Statistics of extremes: modeling ecological disturbances. Ecology 86:1124–1134
Kuhner M (2009) Coalescent genealogy samplers: windows into population history. Trends Ecol Evol 24:86–93
Kuparinen A, Savolainen O, Schurr FM (2010) Increased mortality can promote evolutionary adaptation of forest trees to climate change. For Ecol Manage 259:1003–1008
Kurtz WA, Dymond CC, Rampley GJ, Neilson ET, Carroll AL, Ebata T, Safranyik L (2008) Mountain pine beetle and forest carbon feedback to climate change. Nature 452:987–990
Lake PS (2011) Drought and aquatic ecosystems: effects and responses. Wiley-Blackwell, West Sussex
Larcher W (2000) Temperature stress and survival ability of Mediterranean sclerophyllous plants. Plant Biosyst 134:279–295
Leblanc M, Tweed S, Van Dijk A, Timbal B (2012) A review of historic and future hydrological changes in the Murray-Darling Basin. Global Planet Change 80–81:226–246
Li H, Durbin R (2011) Inference of human population history from individual whole-genome sequences. Nature 475:493–496
Lytle DA, Poff NL (2004) Adaptation to natural flow regimes. Trends Ecol Evol 19:94–100
Marchand FL, Verlinden M, Kockelbergh F, Graae BJ, Beyens L, Nijs I (2006) Disentangling effects of an experimentally imposed extreme temperature event and naturally associated desiccation on Arctic tundra. Funct Ecol 20:917–928
Matusick G, Ruthrof KX, Hardy GSJ (2012) Drought and heat triggers sudden and severe dieback in a dominant mediterranean-type woodland species. Open J Forest 2:183–186
Matusick G, Ruthrof KX, Brouwers NC, Dell B, Hardy GSJ (2013) Sudden forest collapse corresponding with extreme drought and heat in a Mediterranean-type eucalypt forest in southwestern Australia. Eur J For Res 132:497–510
Mayence CE, Marshall DJ, Godfree R (2010) Hydrological and mechanical control for an invasive wetland plant, *Juncus ingens*, and implications for rehabilitating and managing Murray River floodplain wetlands, Australia. Wetl Ecol Manag 18:717–730
Mckeon G, Hall W, Henry B, Stone G, Watson I (eds) (2004) Pasture degradation and recovery in Australia's rangelands: learning from history. Queensland Department of Natural Resources, Mines and Energy, Brisbane, QLD
McMahon TA, Finlayson BL (2003) Droughts and anti-droughts: the low flow hydrology of Australian rivers. Freshw Biol 48:1147–1160
Menzel A, Sparks TH, Estrella N, Koch E, Aasa A, Ahas R, Alm-kübler K, Bissolli P, Braslavská O, Briede A, Chmielewski FM, Crepinsek Z, Curnel Y, Dahl A, Defila C, Donnelly A, Filella Y, Jatczak K, Mage F, Mestre A, Nordli O, Peñuelas J, Pirinen P, Remišová V, Scheifinger H,

Striz M, Susnik A, Van Vliet AJH, Wielgolaski FE, Zach S, Zust A (2006) European phenological response to climate change matches the warming pattern. Glob Chang Biol 12: 1969–1976
Miller W, Schuster SC, Welch AJ, Ratan A, Bedoya-Reina OC, Zhao F, Kim HL, Burhans RC, Drautz DI, Wittekindt NE, Tomsho LP, Ibarra-Laclette E, Herrera-Estrella L, Peacock E, Farley S, Sage GK, Rode K, Obbard M, Montiel R, Bachmann L, Ingólfsson Ó, Aars J, Mailund T, Wiig Ø, Talbot SL, Lindqvist C (2012) Polar and brown bear genomes reveal ancient admixture and demographic footprints of past climate change. Proc Natl Acad Sci 109:2382–2390
Murphy BF, Timbal C (2008) A review of recent climate variability and climate change in southeastern Australia. Int J Climatol 28:859–879
Musil CF, Schmiedel U, Midgley GF (2005) Lethal effects of experimental warming approximating a future climate scenario on southern African quartz-field succulents: a pilot study. New Phytol 165:539–547
Naiman RJ, Latterell JJ, Pettit NE, Olden JD (2008) Flow variability and the biophysical vitality of river ecosystems. Geoscience 340:629–643
Parmesan C, Root TL, Willig MR (2000) Impacts of extreme weather and climate on terrestrial biota. Bull Am Meteorol Soc 81:443–450
Phillips PC (1996) Maintenance of polygenic variation via a migration-selection balance under uniform selection. Evolution 50:1334–1339
Pindyck RS (2011) Fat tails, thin tails, and climate change policy. Rev Environ Econ Policy 5:258–274
Pook EW, Forrester RI (1984) Factors influencing dieback of drought-affected dry sclerophyll forest tree species. Aust For Res 14:201–217
Pook EW, Costin AB, Moore CWE (1966) Water stress in native vegetation during the drought of 1965. Aust J Bot 14:257–267
Pook M, Lisson S, Risbey J, Ummenhofer CC, McIntosh P, Rebbeck M (2009) The autumn break for cropping in southeast Australia: trends, synoptic influences and impacts on wheat yield. Int J Climatol 29:2012–2202
Prentice IC, Cramer W, Harrison SP, Leemans R, Monserud RA, Solomon AM (1992) A global biome model based on plant physiology and dominance, soil properties and climate. J Biogeogr 19:117–134
Reid MA, Ogden RW (2006) Trend, variability or extreme event? The importance of long-term perspectives in river ecology. River Res Appl 22:167–177
Reznick DN, Ghalambor CK (2001) The population ecology of contemporary adaptations: what empirical studies reveal about the conditions that promote adaptive evolution. Genetica 112–113:183–198
Ross-Ilbarra J, Wright SI, Foxe JP, Kawabe A, DeRose-Wilson L, Gos G, Charlesworth D, Gaut BS (2008) Patterns of polymorphism and demographic history in natural populations of Arabidopsis lyrata. PLoS One 3, e2411
Schubert SD, Suarez MJ, Pegion PJ, Koster RD, Bacmeister JT (2004) On the cause of the 1930s dust bowl. Science 303:1855–1859
Sgrò CM, Lowe AJ, Hoffmann AA (2011) Building evolutionary resilience for conserving biodiversity under climate change. Evol Appl 4:326–337
Siol M, Wright SI, Barrett SCH (2010) The population genomics of plant adaptation. New Phytol 188:313–332
Smith MD (2011) An ecological perspective on extreme climatic events: a synthetic definition and framework to guide future research. J Ecol 2011:656–663
Stephenson B (2008) Definition, diagnosis, and origin of extreme weather and climate events. In: Diaz HF, Murnane RJ (eds) Climate extremes and society. Cambridge University Press, Cambridge, pp 1–22
Stoneking M, Krause J (2011) Learning about human population history from ancient and modern genomes. Nat Rev Genet 12:603–614
Stott PA, Stone DA, Allen MR (2004) Human contribution to the European heatwave of 2003. Nature 432:610–613

Tigerstedt PMA (1994) Adaptation, variation and selection in marginal areas. Euphytica 77:171–174

Verdon-Kidd DC, Kiem AS (2009) Nature and causes of protracted droughts in southeast Australia: comparison between the Federation, WWII and Big Dry droughts. Geophys Res Lett 36, L22707

Vivian LM, Godfree RC, Colloff MJ, Mayence CE, Marshall DJ (2014a) Wetland plant growth under contrasting water regimes associated with river regulation and drought: implications for environmental water management. Plant Ecol 215:997–1011

Vivian LM, Ward KA, Zwart AB, Godfree RC (2014b) Environmental water allocations are insufficient to control an invasive plant: evidence from a highly regulated floodplain wetland. J Appl Ecol 51:1292–1303

Walker BH (1993) Rangeland ecology: understanding and managing change. Ambio 22:80–87

Walker KF, Sheldon F, Puckridge JT (1995) A perspective on dryland river ecosystems. Regul Rivers: Res Manage 11:85–104

Walther G-R, Post E, Convery P, Menzel A, Parmesan C, Beebee TJC, Fromentin J-M, Hoegh-Guldberg O, Bairlein F (2002) Ecological responses to recent climate change. Nature 416:389–395

Woodhouse CA, Overpeck JT (1998) 2000 years of drought variability in the central United States. Bull Am Meteorol Soc 79:2693–2714

Zhao M, Running SW (2010) Drought-induced reduction in global terrestrial net primary production from 2000 through 2009. Science 329:940–943

Control of Arable Crop Pathogens; Climate Change Mitigation, Impacts and Adaptation

Bruce D.L. Fitt, David John Hughes, and Henrik Uwe Stotz

Introduction

Crop diseases directly threaten global food security because diseases cause crop losses, estimated at 16 % globally, despite efforts to control the diseases (Fisher et al. 2013; Oerke 2006), in a world where more than one billion people do not have enough food (FAO 2009). Thus, food production must be increased by controlling crop diseases more effectively. Food security problems associated with crop diseases can be exacerbated by climate change (Fitt et al. 2011; Garrett et al. 2006; Gregory et al. 2009). Since the threats of climate change to food security are particularly severe in marginal areas (Schmidhuber and Tubiello 2007), there is pressure on farmers in fertile areas that may benefit from climate change, such as northern Europe (Butterworth et al. 2010), to produce more food to ensure global food security (Stern 2007). Thus, it is essential to include methods to control disease problems in strategies for adaptation to impacts of climate change (Gregory et al. 2009; Evans et al. 2008). However, it is also necessary to grow crops in countries such as the UK in a manner that decreases emissions of greenhouse gases (GHG) to contribute now to climate change mitigation from agriculture (Hughes et al. 2011; Jackson et al. 2007). To decrease the contribution of agriculture to global warming, possible options include decreasing the use of fossil fuels and nitrogen fertilisers, decreasing methane emissions from livestock and increasing the sequestering of carbon from the atmosphere (Glendining et al. 2009; Smith et al. 2008).

This chapter describes three aspects of the interactions between climate change and diseases that reduce arable crop yields:

B.D.L. Fitt, M.A., Ph.D., D.I.C., F.R.S.B. (✉) • H.U. Stotz, Ph.D.
School of Life & Medical Sciences, University of Hertfordshire,
College Lane Campus, Hatfield, Hertfordshire AL10 9AB, UK
e-mail: b.fitt@herts.ac.uk

D.J. Hughes, M.Sc., D.Phil.
Computational & Systems Biology, Rothamsted Research, Harpenden, Hertfordshire, UK

D. Edwards, J. Batley (eds.), *Plant Genomics and Climate Change*,
DOI 10.1007/978-1-4939-3536-9_3

- Climate change mitigation; consequences for greenhouse gas (GHG) emissions of crop management strategies to control diseases, using UK oilseed rape and barley crops as examples.
- Impacts of climate change on incidence of crop diseases and their effects on crop yields, using UK oilseed rape phoma stem canker and wheat fusarium ear blight as examples.
- Adaptation of crop disease management strategies to decrease arable crop losses related to climate change.

Crop Disease Control and Climate Change Mitigation

In 2008, agriculture accounted for 7.7 % of UK GHG emissions (48 Mt CO_2 eq.; DECC 2012) and these were primarily in the non-CO_2 sector. As part of the overall 80 % emissions reduction strategy, the UK Committee on Climate Change has set a target of a 70 % reduction in the non-CO_2 sector by 2050 (Committee on Climate Change 2010). This has prompted debate about how best to decrease GHG whilst maintaining food production. One question in this debate is whether the use of fungicides and other treatments to control crop diseases leads to an increase or decrease in GHG emissions, with the associated environmental consequences.

Mahmuti et al. (2009) calculated the GHG emissions for production of 1 t of winter oilseed rape seed. The differences in yields between fungicide-treated and untreated plots were then analysed to estimate the effects of fungicides on GHG emissions per tonne of seed. This was done for data from UK winter oilseed rape experiments in harvest years 2004–2007 (Fig. 1). The analysis takes account of GHG emissions associated with the manufacture and application of fertilisers and fungicides, and with the field operations of spraying, harvesting, drying etc. The production of 1 ha of winter oilseed rape was estimated to release emissions of 3337 kg CO_2 eq. The GHG emissions per tonne of seed produced decreased as the yield of the seed increased; the difference in GHG emissions per tonne between yields of 1 and 3 t ha^{-1} was 2225 kg CO_2 eq. t^{-1}. In the series of experiments over 4 years, mean yields were 4.33 t ha^{-1} for fungicide-treated crops and 3.84 t ha^{-1} for untreated crops. Thus the disease-induced yield loss of approximately 11.3 % of the fungicide-treated winter oilseed rape yield was associated with a net increase in emissions of 98 kg CO_2 eq. t^{-1} for winter oilseed rape (Fig. 1). Crop yields depend on many factors and vary from year to year but the same results were obtained in a wide range of comparisons. One important factor was the extent to which different cultivars of oilseed rape were susceptible or resistant to pathogens such as *Pyreuopeziza brassicae* or *Leptosphaeria maculans*. Cultivar resistance provides not only the direct benefit of greater yields but also the indirect benefit of reduced GHG emissions per tonne of seed produced. However, such resistance tends to have limited effectiveness against changing virulence in the pathogen population. This raises further questions about how best to deploy resistant cultivars to obtain the most benefit from them.

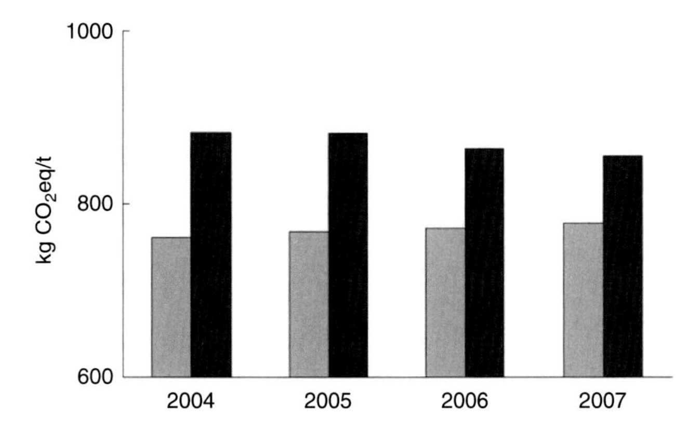

Fig. 1 Differences in greenhouse gas (GHG) emissions per tonne of yield between winter oilseed rape crops (means of 24–39 cultivars at 4–7 different sites) treated with fungicides to control phoma stem canker and light leaf spot diseases (*light grey region*) and untreated crops (*black region*) in HGCA field experiments), at sites differing in epidemic severity. The numbers of sites where the data were available for both treated and untreated crops were 5 (2004), 7 (2005), 6 (2006) and 4 (2007). The numbers of cultivars used in different years were 26 (2004), 39 (2005), 24 (2006) and 29 (2007). [Adapted from Mahmuti M, West JS, Watts J, Gladders P, Fitt BDL. Controlling crop disease contributes to both food security and climate change mitigation. Int J Agric Sust. 2009; 7: 189–202. With permission Taylor & Francis]

Hughes et al. (2011) did similar calculations for 28 cultivars of winter and spring barley, grown at 24 UK sites. The inputs to growing winter barley (including fungicides) were estimated to release emissions of 2617 kg CO_2 eq. ha^{-1}. The corresponding emissions for spring barley were 20 % less (2099 kg CO_2 eq. ha^{-1}), mainly because nitrogen inputs were smaller. These estimates are smaller than comparable estimates for oilseed rape (Mahmuti et al. 2009) and wheat (Berry et al. 2008), mainly because average rates of fertiliser application to barley crops are smaller. Across all datasets, fungicide treatment reduced GHG emissions by 42–60 kg CO_2 eq. t^{-1} (11–16 %) for winter barley and by 29–39 kg CO_2 eq. t^{-1} (8–11 %) for spring barley. The reductions in GHG emissions were larger when fungicide treatment was more effective in increasing yields. In addition, the decrease in GHG emissions was generally greater for winter barley than for spring barley, because winter barley production emits more GHGs than spring barley production. There were reductions in GHG emissions across a wide range of comparisons. A sensitivity analysis confirmed that disease control continues to give reductions in GHG emissions, even if alternative, substantially greater values were used for the emission factors associated with agricultural pesticides.

Combining the decreases in GHG emissions associated with disease control in UK winter wheat, winter oilseed rape, winter barley and spring barley crops, Hughes et al. (2011) estimated that for the UK such disease control in arable crops decreased GHG by c. 1.6 Mt CO_2eq. each year from 2005 to 2009 (Fig. 2), making a substantial contribution to government targets for decreasing GHG associated with agriculture.

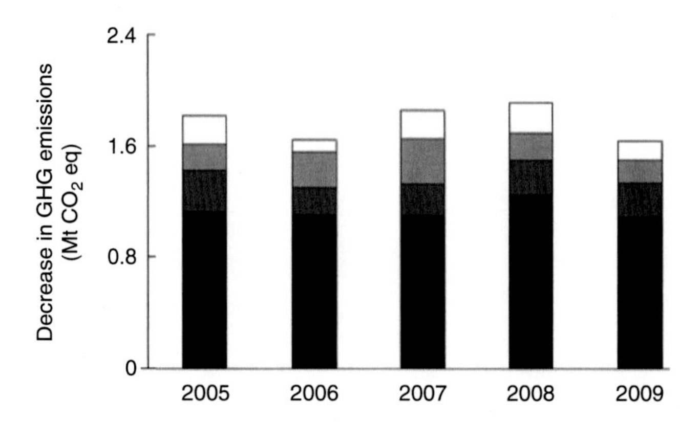

Fig. 2 Estimated decrease in GHG emissions (Mt CO2 eq.) through use of fungicides to control diseases and increase yields in winter wheat (*black region*), winter oil seed rape (*dark grey region*) winter barley (*light grey region*) and spring barley (*white region*) for the United Kingdom in harvest years 2005–2009. Total decreases in GHG emissions are 15 % (2005), 14 % (2006), 15 % (2007), 14 % (2008) and 13 % (2009) of the estimated total GHG emissions (Mt CO_2 eq.) if these four crops were grown without fungicide treatment. [Adapted from Hughes DJ, West JS, Atkins SD, Gladders P, Jeger MJ, Fitt BDL. Effects of disease control by fungicides on Greenhouse Gas (GHG) emissions by UK arable crop production. Pest Manag Sci. 2011; 67: 1082–1092. With permission from John Wiley & Sons]

There are also more general effects to consider. Suppose arable crop yields were to decrease in future. This could happen for a number of reasons, including climate change, the arrival of new crop diseases in UK, greater susceptibility to existing diseases or reductions in the types or quantities of permitted fungicides. To maintain UK production at the same level as today, more arable land would be needed. Mahmuti et al. (2009) estimated that an additional 680,000 ha would be required to sustain UK agricultural production at 2007 levels, if crops were untreated rather than sprayed with fungicides. This area represents land that could otherwise be used to grow more food or biofuel crops, or as a wildlife habitat. Furthermore, since both biomass and soil organic carbon are typically released when uncultivated land is cultivated, use of such land for crops can increase GHG emissions. Taking 200 t CO_2 eq. ha^{-1} as an estimate of the GHGs emitted by converting temperate grassland into arable crop land, the conversion of 688,000 ha UK grassland to agricultural crops would release more than 100 Mt CO_2.

Generally speaking, measures that decrease crop yields (e.g. reduced use of fungicides and N fertilisers) are likely to result in an overall increase in GHG emissions, due to the necessary expansion of land under cultivation (Berry et al. 2008; Burney et al. 2010; Carlton et al. 2012). On the other hand, certain soil fertility management practices associated with organic crop farming have the potential to sequester soil organic carbon (SOC) in long-term arable land (Azeez 2009) and a reduction in inputs can decrease emissions (Lin et al. 2011).

Carlton et al. (2012) compared the annual GHG emissions from UK arable production under the conventional (current) crop production system with the emissions predicted assuming the nationwide adoption of reduced tillage, organic or integrated arable systems whilst maintaining current crop production. The ‘reduced tillage’ arable system is similar to conventional crop production except that reduced tillage methods are adopted wherever practical (not usually for crops requiring considerable soil cultivation, such as potatoes, sugar beet etc.). The ‘organic’ arable system assumes that there are no applications of synthetic crop protection products or synthetic fertilisers, and that crop rotations include fertility-enhancing periods. The ‘integrated’ crop production system integrates the high yields of conventional crop production with the SOC sequestration of organic crop production, employing fertility enhancing rotations and use of organic manure to augment soil organic carbon, but allowing use of additional synthetic fertilisers, fungicides for disease control and other crop protection products to achieve conventional crop yields.

This analysis suggests that conventional farming, plus reduced tillage cultivation where appropriate, can best contribute to the achievement of government GHG emissions targets. The reduced tillage system demonstrated a modest (<20 %) reduction in emissions in all cases, although in practice it may not be suitable for all soils and is likely to cause problems with control of diseases spread on crop debris. However, there were substantial increases in GHG emissions associated with the organic and integrated systems nationally, principally due to soil organic carbon losses from land use change. The integrated system includes a 50 % fertility-enhancing rotation, which increases the total average UK arable area (currently about 6 Mha) by 2.7 Mha. The area of arable land under the organic system would be more than double the arable area under the conventional system through the combined impacts of smaller yields and the 50 % fertility-enhancing rotation.

It is important to recognise that local or regional factors can greatly affect the conclusions from studies such as these. For example, soil water losses are likely to become increasingly important in the south and east of England, where climate change is predicted to lead to drier summers. Reduced tillage can reduce such losses, but can also increase the severity of disease epidemics caused by residual inoculum on crop debris. This is a small example of a general conclusion from these studies, that the interaction between climate change, GHG emissions, crop diseases and agricultural production is very complicated. In addition, all crop management systems must operate within cultural and economic constraints.

Impacts of Climate Change on Crop Diseases

There is a need to evaluate impacts of climate change on disease-induced losses in crop yield to guide government and industry policy and planning for adaptation to climate change. It is essential to identify those current crop diseases that may increase in severity or range and those pathogens that may spread to new areas. Impacts of climate change on crop yields may be especially severe in developing countries, where food security problems are already most acute because diseases can destroy crops and cause famine for subsistence farming families, who have few alternatives to their staple crops (Strange and Scott 2005). There is an urgent need to identify potential impacts of climate change on crop diseases now because it can take 10–15 years to breed a new crop cultivar or develop a new fungicide (Fitt et al. 2011) and implementation of policy changes in agriculture also takes time since farmers are often reluctant to change long-established practices.

Methods to assess potential impacts of climate change on crop disease-induced losses have improved greatly over the last few years. Early attempts to assess such impacts used qualitative, rule-based reasoning that could not easily accommodate the complex host-pathogen-environment interactions involved (Coakley et al. 1999; Anderson et al. 2004). This work did not use simulated weather generated by general circulation models but relied on predictions of fixed changes in temperature and rainfall. Little of this work was based on data that was extensive enough to allow use of separate independent data sets for model development and model validation, respectively. This work did not always clearly distinguish between direct impacts and indirect impacts of climate change on crop diseases. Indirect impacts of climate change on crops are extremely difficult to assess, let alone to model. For example, increasing temperature in the UK may have contributed to the increase in the area of maize grown, which may in turn have contributed to the increase in incidence of the mycotoxin-producing *Fusarium graminearum*, since maize debris is a potent source of inoculum of this pathogen (West et al. 2012a).

However, direct impacts of changes in weather patterns as a result of climate change can be modelled more easily. General circulation models were used as a basis for projections of an increase in the range of *Phytophthora cinnamomi* disease on oak trees in France (Bergot et al. 2004). Intergovernmental Panel on Climate Change (IPCC) global high and low CO_2 emission scenarios (Nakicenovic et al. 2000) were used as a basis for UKCIP02 climate change projections for the 2020s and 2050s, using regional climate models, by comparison with a baseline period (1960–1990) (Hulme et al. 2002). UKCIP02 provides predicted changes in monthly climate variables on a 50 km grid. The LARS-WG weather generator used UKCIP02 projections to produce yearly site-specific daily weather for the 2020s and 2050s (Semenov 2007). Simulated daily weather for 70 years was used to project an increase in the range and severity of phoma stem canker on UK oilseed rape (Evans et al. 2008). Later work has been able to use new IPPC climate change scenarios; for work on projections of fusarium ear blight in China, the A1B climate change scenario was used (Zhang et al. 2014). It is important to assess the impacts of climate change on both crop growth and the disease epidemiology; to ignore one of them can produce inaccurate projections (Butterworth et al. 2010; Madgwick et al. 2011).

There are a number of steps required to assess such impacts (Fig. 3; Madgwick et al. 2011). Firstly, there is a need to assemble a good set of observed crop growth, disease incidence/severity and weather data. Generally, at least 10 years of data from a range of sites in the region of interest is required. These data can then be divided into two parts. Two thirds of the data can be used for model construction and the remainder used to provide an independent data set for model validation; it is important that both data sets span the range of sites and years. In the case of fusarium

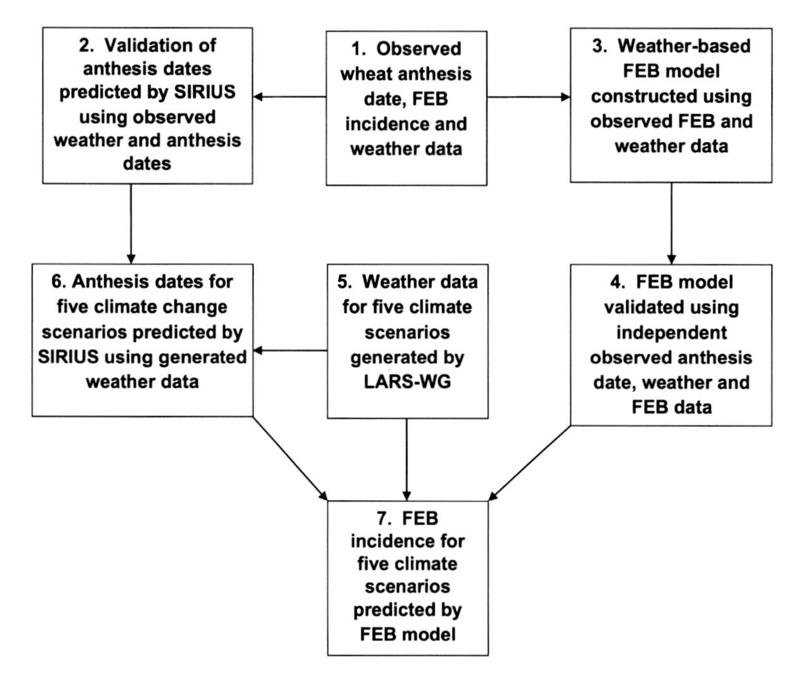

Fig. 3 An illustration of how the different models were combined to produce projections of date of winter wheat anthesis (growth stage 65) and fusarium ear blight (FEB) incidence (% plants affected) for different climate change scenarios. (*1*) Observed data for weather (daily minimum and maximum temperature (°C), total rainfall (mm) and solar radiation (MJ day^{-1})), date of anthesis and fusarium ear blight incidence were collated from a number of sources for different regions of the UK for the years 1994–2008. (*2*) The dates of anthesis predicted using the wheat growth model SIRIUS were validated by comparing predicted anthesis dates for winter wheat cv. Consort, generated by SIRIUS using observed weather data, with observed anthesis dates for the same sites for the period 1997–2004. (*3*) A fusarium ear blight model was developed from data for fusarium ear blight incidence from sites within 80 km of Rothamsted and observed weather for Rothamsted for the period 1994–2008; the model related fusarium ear blight incidence to average May temperature and rainfall in the second week of June (time of observed anthesis dates for Rothamsted). (*4*) Predictions of average percentage of plants affected by fusarium ear blight were validated by comparing predictions made using observed weather to observed fusarium ear blight incidence data for the period 1994–2008 for different regions of the UK (northeast, southwest and east England) which were plotted as north (northeast) and south (southwest and east) England on the validation graph. (*5*) Weather data were generated using LARS-WG for each of the 14 sites for each climate scenario; baseline (based on the statistical variability (or patterns) in observed weather variables in the period 1960–1990) and high CO_2 and low CO_2 emissions scenarios for the 2020s and 2050s (2020LO, 2020HI, 2050LO and 2050HI). (*6*) The dates of anthesis for cv. Consort were projected for each site for each climate scenario using SIRIUS, allowing maps to be generated to show the effect of climate change on date of anthesis. (*7*) Using the weather generated by LARS-WG and average date of anthesis projected using SIRIUS for each of the sites for each of the five climate scenarios, the fusarium ear blight model was used to project fusarium ear blight incidence for each site for each of the five climate scenarios. [Adapted from Madgwick JW, West JS, White RP, Semenov MA, Townsend JA, Turner JA, et al. Impacts of climate change on wheat anthesis and fusarium ear blight in the UK. Eur J Plant Pathol. 2011; 130: 117–131. With permission from Springer Science]

ear blight, the data required were the date of anthesis (since the crop is susceptible only at this growth stage (Xu and Nicholson 2009; Xu et al. 2007)), incidence of fusarium ear blight, temperature and rainfall.

Having assembled the data, it is then necessary to produce weather-based crop growth and disease incidence/severity models. Such models need to be simple and should not include parameters for which simulated weather associated with different climate change scenarios cannot be generated. There may be an existing crop growth model that can be calibrated for the region of interest; for example the SIRIUS wheat growth model (Jamieson et al. 1998) was used for work on fusarium ear blight (Madgwick et al. 2011) and the STICS oilseed rape growth model (Brisson et al. 2003) was used for work on phoma stem canker (Butterworth et al. 2010); the STICS model was developed in France but the radiation use efficiency parameter was modified so that the model fitted oilseed rape yields in the UK. It is frequently the case that weather-based disease models developed elsewhere do not fit the region of interest and new region-specific models have to be developed (Zhang et al. 2014; Madgwick et al. 2011). When these weather-based crop growth and disease incidence/severity models have been developed, it is necessary to validate them with independent data.

In parallel, it is necessary to produce simulated weather data for the region and climate change scenario of interest. In UK work with phoma stem canker of oilseed rape and fusarium ear blight of wheat, these weather data were provided by the LARS-WG stochastic weather generator (Semenov 2007), whereas in the Chinese work with fusarium ear blight (Zhang et al. 2014) they were provided by PRECIS (Jones et al. 2004). These simulated weather data can then be used as inputs into the crop and disease models to estimate the impacts of climate change on the disease at sites in the specific region for the selected climate change scenario (Evans et al. 2008; Zhang et al. 2014; Madgwick et al. 2011). Whilst the outputs of such assessments of impacts of climate change on crop diseases are generated for specific sites, they can be converted into maps by spatial interpolation between those sites (Fig. 4). Thus it was projected that climate change will increase incidence of wheat fusarium ear blight and severity of phoma stem canker in the UK, especially in Southern England (Evans et al. 2008; Madgwick et al. 2011).

By using yield loss formulae relating yield loss (t/ha) to incidence or severity of a specific disease, these data then can be used to estimate the losses associated with diseases under different climate change scenarios. For example, Butterworth et al. (2010) combined the STICS oilseed rape crop growth model with simulated weather to project that yields of oilseed rape in which diseases are controlled will increase in the 2020s and 2050s under both high and low CO_2 emission scenarios, especially in Scotland but also in some regions of England (Table 1). By contrast, when phoma stem canker was not controlled, there was a projected decrease in yield, especially in southern England. Subsequently, further work included light leaf spot disease, which is generally most severe in Scotland, by comparison with phoma stem canker, which is generally most severe in southern England. When crop prices were added, it was possible to estimate impacts of climate on values of crop disease losses in different regions of England and Scotland (Tables 2 and 3 (Evans et al. 2010)).

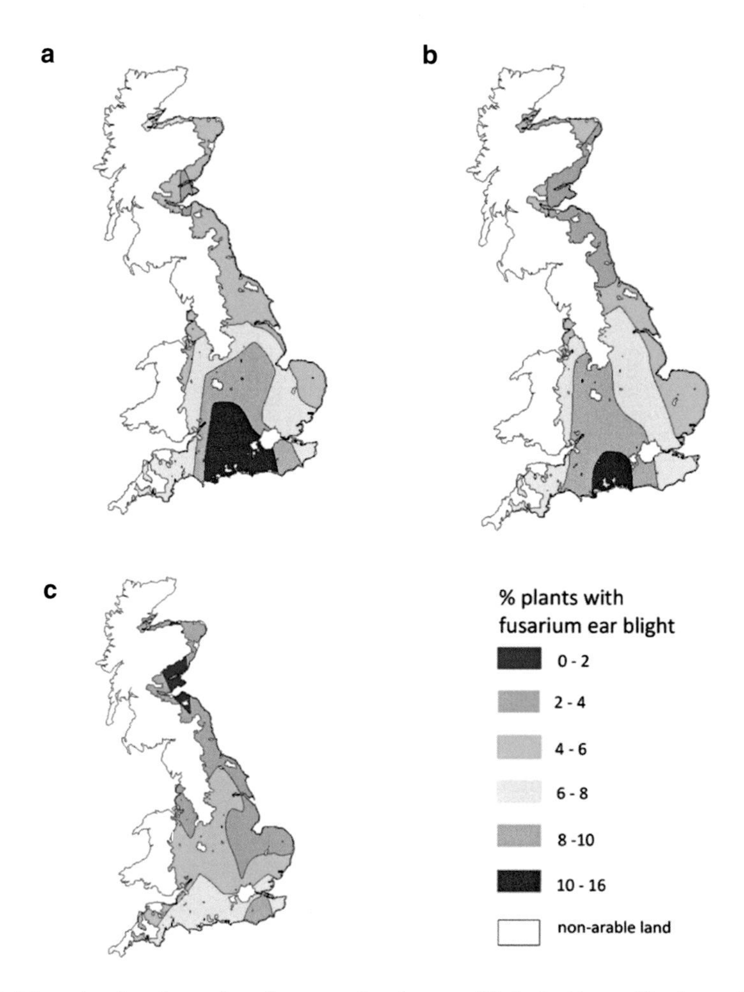

Fig. 4 Maps showing the projected average fusarium ear blight incidence (% plants affected) generated by the fusarium ear blight model using the estimated average anthesis dates for three climate change scenarios; (**a**) baseline, (**b**) 2050LO and (**c**) 2050HI. The baseline scenario is based on the patterns in observed weather from 1960 to 1990, and the other scenarios are high CO_2 (HI) and low CO_2 (LO) emissions scenarios for the 2050s. The maps were produced by spatial interpolation between 14 sites of weather stations distributed across the arable area of the UK

Such projections can help to guide forward financial planning, although they involve various assumptions about changes in prices with time. Such assumptions must be clearly stated.

Whilst these projections apply to specific diseases, it is also possible to classify diseases into groups according to specific aspects of their pathogen life cycle in relation to climate change projections (West et al. 2012b). It is important to realise that there are uncertainties in such projections, associated with uncertainties in the projected weather, crop growth or disease models. Nevertheless, such projections

Table 1 Effects of climate change on the yield of treated oilseed rape (Tr) and untreated oilseed rape (Unt) after phoma stem canker losses, calculated by region[a]

Region[b]	Area oilseed rape (ha)[c]	Baseline yield (t/ha)		Yield (% of baseline yield)							
				2020LO		2020HI		2050LO		2050HI	
		Tr	Unt	Tr	Unt	Tr	Unt	Tr	Unt	Tr	Unt
North East	22,787	3.16	2.78	93.4	90.1	103.1	98.3	103.9	96.5	105.1	93.3
North West	3601	2.98	2.48	96.5	92.5	88.7	84.2	100.9	92.4	103.4	89.8
Yorks and Humberside	61,068	3.12	2.64	95.0	90.7	102.8	97.3	102.4	93.8	103.1	89.3
East Midlands	113,479	3.11	2.59	100.7	95.2	100.4	94.0	101.1	91.1	102.7	86.9
West Midlands	34,419	3.00	2.37	99.6	94.2	83.4	78.2	103.5	94.0	107.6	91.4
Eastern	103,488	3.16	2.58	100.0	94.5	99.7	93.1	103.0	92.8	104.7	88.3
London and South East	79,063	3.01	2.34	100.8	95.4	100.9	94.4	103.7	93.0	106.9	89.1
South West	44,858	3.05	2.41	100.3	95.1	100.5	94.2	103.1	93.7	106.7	90.7
England total	462,764	3.09	2.52	99.3	94.1	99.5	93.4	102.6	92.9	104.8	88.9
Scotland	35,780	3.15	3.06	104.8	103.2	107.1	105.0	109.7	96.9	111.5	103.6
UK total	498,544	3.12	2.77	101.8	98.7	103.0	99.3	105.9	94.9	107.9	96.4

[a]The untreated oilseed rape was calculated as the mean of susceptible and resistant cultivars. The area grown per region (2006) and the predicted average regional yield are given for the baseline (1960–1990) scenario. The predicted regional yield as a percentage of the baseline scenario is given for the 2020LO (low CO_2 emission), 2020HI (high CO_2 emission), 2050LO and 2050HI climate scenarios. The figures were calculated after interpolating the results from the treated oilseed rape yield predictions and the stem canker yield loss predictions according to UK government region. [Based on data from ref. Butterworth et al. (2010), with corrected data for Scotland and UK total]

[b]Government regions can be found at http://www.statistics.gov.uk/geography/downloads/uk_gor_cty.pdf

[c]Area of winter oilseed rape grown in each region in harvest year 2006 (www.defra.gov.uk)

Table 2 Effects of climate change on the output of winter oilseed rape (treated with fungicide), calculated by region[a]

	Value of oilseed rape crop (£000s)[c]				
Region[b]	Baseline	2020LO	2020HI	2050LO	2050HI
North East	14,098	13,168	14,536	14,646	14,812
North West	2097	2024	1861	2115	2169
Yorkshire and Humberside	37,220	35,342	38,251	38,126	38,358
East Midlands	69,007	69,480	69,277	69,744	70,874
West Midlands	20,194	20,121	16,839	20,900	21,726
Eastern	63,885	63,854	63,661	65,792	66,907
London and South East	46,508	46,867	46,939	48,216	49,700
South West	26,742	26,831	26,873	27,570	28,538
England total	279,749	277,688	278,237	287,110	293,085
Scotland	22,038	23,086	23,600	24,182	24,567
UK total	301,787	300,774	301,837	311,292	317,652

[a]The area grown per region (2006) and the predicted regional output are given for the baseline (1960–1990), 2020LO (low CO_2 emissions), 2020HI (high emissions), 2050LO and 2050HI climate scenarios and presented in thousands of pounds (£000s). The yield figures were calculated after interpolating the results from the oilseed rape yield predictions according to UK government region and then multiplied by an average price of £195.60 t^{-1}. [Adapted from Evans N, Butterworth MH, Baierl A, Semenov MA, West JS, Barnes A, et al. The impact of climate change on disease constraints on production of oilseed rape. Food Security. 2010; 2: 143–156. With permission from Springer Science]
[b]Government regions can be found at http://www.statistics.gov.uk/geography/downloads/uk_gor_cty.pdf
[c]Corrected data for Scotland and UK total

are widely appreciated by politicians who have to make long-term policy decisions based on the best available projections. There has never been a time when guidance on climate change and crop diseases is more clearly appreciated to guide strategies for adaptation to climate change.

Adaptation of Crop Disease Control to Climate Change

Various adaptation strategies are available to minimise or negate predicted climate change related increases in yield loss from phoma stem canker in UK winter oilseed rape production (Barnes et al. 2010). A number of forecasts for crop yield, national production and subsequent economic values are presented, providing estimates of impacts on both yield and value for different types of adaptation. Under future climate change scenarios, there will be increasing pressure to maintain or increase crop yields. Losses can be minimised in the short term (up to 2020) with an

Table 3 Effects of climate change on losses from phoma stem canker and light leaf spot (for cultivars with average resistance) in winter oilseed rape crops not treated with fungicide[a]

Region[b]	Value of losses caused by phoma stem canker and light leaf spot (£000s)[c]				
	Baseline	2020LO	2020HI	2050LO	2050HI
North East	3431	3526	3934	4208	4630
North West	520	533	501	602	676
Yorks and Humberside	7804	8118	9074	9661	10,874
East Midlands	15,116	16,869	17,567	18,871	21,748
West Midlands	5038	5539	4716	6244	7308
Eastern	14,481	16,179	16,582	18,454	21,359
London and South East	12,388	13,540	13,874	15,381	17,882
South West	7910	8198	8337	8996	10,191
England total	66,690	72,502	74,584	82,417	94,668
Scotland	7109	7663	7901	10,240	9067
UK total	73,890	80,165	82,485	92,657	103,735

[a]Values are given for the baseline (1960–1990), 2020LO (low CO_2 emissions), 2020HI (high emissions), 2050LO and 2050HI climate scenarios and presented in thousands of pounds (£000s). Figures were calculated after interpolating results from stem canker and light leaf spot yield loss predictions according to UK government region and then multiplied by an average price of £195.60 t^{-1}. [Adapted from Evans N, Butterworth MH, Baierl A, Semenov MA, West JS, Barnes A, et al. The impact of climate change on disease constraints on production of oilseed rape. Food Security. 2010; 2: 143–156. With permission from Springer Science]
[b]Government regions can be found at http://www.statistics.gov.uk/geography/downloads/uk_gor_cty.pdf
[c]The stem canker and light leaf spot loss predictions depend on the crop yield predictions in Table 2 of Evans et al. (2010). With corrected data for Scotland and UK total

autonomous adaptation strategy, which essentially requires some farmer-led changes towards best management practices. However, the predicted impacts of climate change can be negated and, in most cases, improved upon, with planned adaptation strategies. This requires increased funding from both public and private sectors and more directed efforts at adaptation from the producer. Most literature on adaptation to climate change has been conceptual with little quantification of impacts. Quantifying the impacts of adaptation is essential to provide clearer information to guide policy and industry approaches to mitigate future climate change risk.

As indicated, adaptation can be autonomous (i.e. without a conscious strategy) or planned and implemented by the public sector. Farmers may adopt autonomous adaptation to optimise their return on investment (Fig. 5). This adaptation may include less frequent use of oilseed rape crops in rotations to reduce the incidence of phoma stem canker and increase yield. Similarly, cultivar choice and the timing of sowing seeds can be optimised to increase yield. Furthermore, improved timing and frequency of fungicide applications will help to increase crop yields in the short term. However, planned adaptation may be beneficial to mitigate impacts of climate

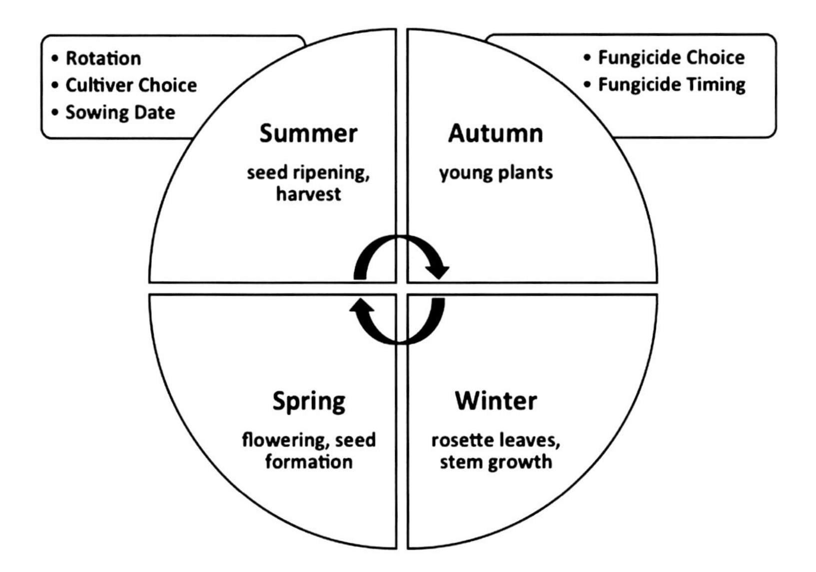

Fig. 5 Seasonal development of winter oilseed rape in the UK in relation to farmer-led autonomous adaptation strategies. Seeds are sown in late summer, rosettes develop in late autumn and stem extension occurs in late winter, followed by flowering in the spring and harvest in the summer. Farmer-led adaptation strategies include crop rotation, cultivar choice (based on HGCA recommended lists) and sowing date to optimize yield. Farmers also optimize the timing and frequency of fungicide sprays based on forecasting (www.rothamsted.ac.uk/phoma-leaf-spot-forecast). [Modified from Barnes AP, Wreford A, Butterworth MH, Semenov MA, Moran D, Evans N, et al. Adaptation to increasing severity of phoma stem canker on winter oilseed rape in the UK under climate change. J Agr Sci 2010; 148: 683–694. With permission from Cambridge University Press]

change on disease-related yield losses. Investment from the private sector should make more effective fungicides for use on oilseed rape available by the 2020s. Moreover, public and private investment, to exploit recent development in our understanding of both host and pathogen genomics, should produce new cultivars with more durable disease resistance. Both advances are expected to increase crop yield and contribute to global food security. The impacts of these two adaptation strategies were modelled to guide government policy and strategic industrial decision making (Barnes et al. 2010).

The yield of winter oilseed rape infected by the phoma stem canker pathogen *Leptosphaeria maculans* was assessed (Fig. 6). Yields were predicted to decrease during the period from the 2020s to the 2050s because occurrence of *L. maculans* is projected to increase under climate scenarios with low or high CO_2 emissions (Butterworth et al. 2010). However, autonomous adaptation involving adopting the best management practices will result in increased yields, even in the short term. Even greater yields are possible in the short term if planned adaptation strategies are also adopted. Adoption of a planned adaptation strategy would benefit the industry in the whole of the UK, although yield increases would be greater in England than in Scotland by the 2050s.

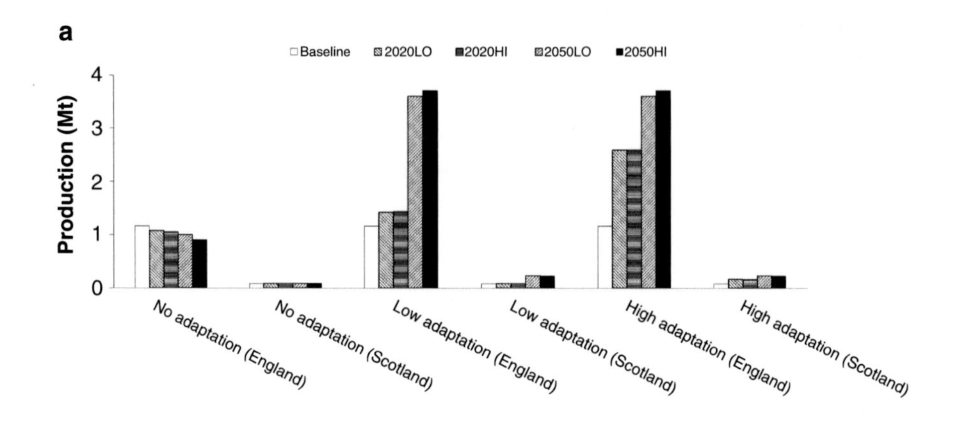

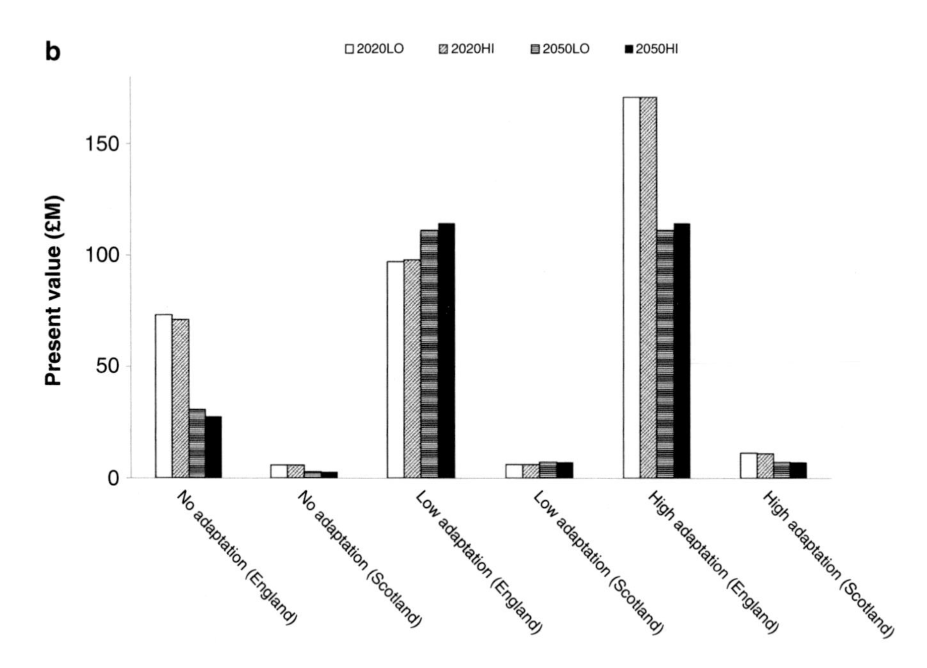

Fig. 6 Impacts of different adaptation strategies on total production (**a**) and present value (**b**) of winter oilseed rape under different climate change scenarios. Impacts of no adaptation, low adaptation or high adaptation were determined. Crops grown in England and Wales (England) or Scotland were considered. High (HI) or low (LO) CO_2 emission scenarios were compared. (Based on data from Barnes et al. (2010))

The economic benefit of different adaptation strategies was calculated, based on the predicted yield responses (Fig. 6 (Barnes et al. 2010)). The value of the crop was based on an average price for the period from 2002 to 2008. Future values were discounted using UK Treasury recommended discount factors of 3.5 and 3 % for

2020 and 2050, respectively. Without adaptation, the value of the crop was expected to decrease from 2020 to 2050. For England and, to a lesser extent, for Scotland, the economic benefit of planned adaptation would be particularly good in the short term. The benefits of adopting a planned adaptation strategy would therefore greatly outweigh the expenditure on research and knowledge transfer by 2020. Interestingly, crop value would still increase until 2050 when autonomous adaptation is implemented, whereas the financial benefits decrease in the case of planned adaptation although this decrease is less than in the absence of adaptation. Discounting may therefore affect the value more when planned adaptation is used. Over the whole period, however, planned adaptation performs better than autonomous adaptation.

Experimental work confirms that an increase in temperature, which is projected under all climate change scenarios, will increase phoma stem canker severity (Huang et al. 2006). An important contributor to the increased susceptibility of oilseed rape cultivars to *L. maculans* at increased temperature is the temperature sensitivity of certain resistance (*R*) genes (e.g. *Rlm6* (Huang et al. 2006)). The propensity of certain genes for resistance against *L. maculans* to become ineffective at elevated temperatures can place a burden on breeders to develop disease resistance in cultivars that is both environmentally stable and durable. Certainly, the use of genetic backgrounds with quantitative resistance will be an important component of breeding programmes (Brun et al. 2010).

Conclusions

These results show that disease control in arable crops can contribute to both climate change mitigation and global food security. They suggest that disease control should be included in policy options for decreasing GHG emissions from agriculture (Smith et al. 2008). Thus, controlling diseases in UK winter oilseed rape and barley gives benefits in terms of decreased GHG per tonne of crop produced and increased yield to contribute to food production in northern Europe in response to climate change threats to global food security (Stern 2007). These decreases in GHG are especially associated with more efficient use of nitrogen fertiliser applied to the crop (Glendining et al. 2009). Furthermore, the climate change mitigation benefits associated with disease control in UK winter oilseed rape are relatively greater than those associated with disease control in winter wheat (Berry et al. 2008) or winter or spring barley (Hughes et al. 2011). It is also likely that there will be climate change mitigation benefits from disease control in other arable crops in different regions of the world.

These results with diseases of UK oilseed rape demonstrate how climate change can increase losses from crop diseases. For UK winter oilseed rape, the increase in losses is associated with the increase in range and severity of phoma stem canker with global warming (Butterworth et al. 2010; Evans et al. 2008). Predicted losses from canker are substantial even though they may be offset by decreasing losses from light leaf spot. This work illustrates how, worldwide, increased disease losses may be associated with increases in severity of existing diseases or spread of diseases to new areas to threaten crop production (Garrett et al. 2006; Gregory et al. 2009). Thus, there is a risk that the 16 % of crop production lost to diseases (Oerke 2006) may increase, with serious consequences for the one billion people who do not have enough to eat (FAO 2009; Strange and Scott 2005), unless appropriate strategies for adaptation to this effect of climate change are put in place. To guide government and industry strategies for adaptation to climate change, there is an urgent need for reliable predictions of impacts of climate change on different diseases, obtained by combining impacts on crop growth and on disease epidemics with predicted future weather patterns (Barnes et al. 2010). Since it may take 10–15 years to develop a new fungicide or incorporate resistance to a crop pathogen from a novel source of resistance, it is important to identify future target diseases now.

In a world where climate change is exacerbating the food security problems for communities farming in marginal environments (Schmidhuber and Tubiello 2007), it is essential to develop better strategies for controlling crop diseases as a contribution to global food security. There is an urgent need to decrease current global average crop losses to diseases from 16 % (Oerke 2006), especially since disease losses are often much greater in crops grown by subsistence farmers in marginal areas. It is environmentally preferable to increase food production by decreasing losses to diseases rather than by expanding the area cultivated with crops, which will lead to destruction of rainforests and other natural ecosystems and increases in GHG emissions. Disease resistance breeding, fungicides and cultural methods can all contribute to strategies to decrease disease losses but they need to be carefully integrated into disease management strategies appropriate for the relevant farming system. There is a need to optimise disease control to maximise crop production in northern Europe both as a contribution to global food security in the face of climate change (Stern 2007) and to maintain the yields and profitability of European farms and thus provide food security for their farming families.

Acknowledgements The authors thank the UK Biotechnology and Biological Sciences Research Council (BBSRC, Bioenergy and Climate Change ISPG) and Department for Environment, Food and Rural Affairs (Defra, OREGIN, IF0144), HGCA and the Sustainable Arable LINK programme (PASSWORD, LK0944; CORDISOR, LK0956; CLIMDIS, LK09111) for funding this research, and the British Society for Plant Pathology for supplementary funding. We thank Michael Butterworth, Neal Evans, James Madgwick, Martin Mahmuti, Mikhail Semenov, Rodger White, Xu Zhang (University of Hertfordshire), Andreas Baierl (University of Vienna), Andrew Barnes, Dominic Moran (SRUC), Rob Carlton (Carlton Consultancy) and Jack Watts (HGCA) for their contributions to this work. We are grateful to many colleagues for supplying data or advice for this work.

References

Anderson PK, Cunningham AA, Patel NG, Morales FJ, Epstein PR, Daszak P (2004) Emerging infectious diseases of plants: pathogen pollution, climate change and agrotechnology drivers. Trends Ecol Evol 19:535–544

Azeez G (2009) Soil carbon and organic farming. A review on the relationship between agriculture and soil carbon sequestration, and how organic farming can contribute to climate change mitigation and adaptation. Soil Association, Bristol

Barnes AP, Wreford A, Butterworth MH, Semenov MA, Moran D, Evans N et al (2010) Adaptation to increasing severity of phoma stem canker on winter oilseed rape in the UK under climate change. J Agric Sci 148:683–694

Bergot M, Cloppet E, Perarnaud V, Deque M, Marcais B, Desprez-Loustau ML (2004) Simulation of potential range expansion of oak disease caused by *Phytophthora cinnamomi* under climate change. Glob Change Biol 10:1539–1552

Berry PM, Kindred DR, Paveley ND (2008) Quantifying the effects of fungicides and disease resistance on greenhouse gas emissions associated with wheat production. Plant Pathol 57:1000–1008

Brisson N, Gary C, Justes E, Roche R, Mary B, Ripoche D et al (2003) An overview of the crop model STICS. Eur J Agron 18:309–332

Brun H, Chevre AM, Fitt BDL, Powers S, Besnard AL, Ermel M et al (2010) Quantitative resistance increases the durability of qualitative resistance to *Leptosphaeria maculans* in *Brassica napus*. New Phytol 185:285–299

Burney JA, Davis SJ, Lobell DB (2010) Greenhouse gas mitigation by agricultural intensification. Proc Natl Acad Sci U S A 107:12052–12057

Butterworth MH, Semenov MA, Barnes A, Moran D, West JS, Fitt BDL (2010) North-south divide; contrasting impacts of climate change on crop yields in Scotland and England. J R Soc Interface 7:123–130

Carlton RR, West JS, Smith P, Fitt BDL (2012) A comparison of GHG emissions from UK field crop production under selected arable systems with reference to disease control. Eur J Plant Pathol 133:333–351

Coakley SM, Scherm H, Chakraborty S (1999) Climate change and plant disease management. Annu Rev Phytopathol 37:399–426

Committee on Climate Change (2010) Meeting carbon budgets – ensuring a low-carbon recovery. Second progress report to parliament. http://downloads.theccc.org.uk/0610/pr_meeting_carbon_budgets_full_report.pdf

DECC (2012) Agriculture GHG inventory summary factsheet. Department of Energy and Climate Change, London, https://www.gov.uk/government/publications/greenhouse-gas-inventory-summary

Evans N, Baierl A, Semenov MA, Gladders P, Fitt BDL (2008) Range and severity of plant disease increased by global warming. J R Soc Interface 5:525–531

Evans N, Butterworth MH, Baierl A, Semenov MA, West JS, Barnes A et al (2010) The impact of climate change on disease constraints on production of oilseed rape. Food Secur 2:143–156

FAO (2009) 1.02 Billion people hungry; one sixth of humanity undernourished – more than ever before. FAO (Food and Agriculture Organisation of the United Nations). http://www.fao.org/news/story/en/item/20568/icode/

Fisher MC, Henk DA, Briggs CJ, Brownstein JS, Madoff LC, McCraw SL et al (2013) Emerging fungal threats to animal, plant and ecosystem health. Nature 484:186–194

Fitt BDL, Fraaije BA, Chandramohan P, Shaw MW (2011) Impacts of changing air composition on severity of arable crop disease epidemics. Plant Pathol 60:44–53

Garrett KA, Dendy SP, Frank EE, Rouse MN, Travers SE (2006) Climate change effects on plant disease: genomes to ecosystems. Annu Rev Phytopathol 44:489–509

Glendining MJ, Dailey AG, Williams AG, van Evert FK, Goulding KWT, Whitmore AP (2009) Is it possible to increase the sustainability of arable and ruminant agriculture by reducing inputs? Agr Syst 99:117–125

Gregory PJ, Johnson SN, Newton AC, Ingram JS (2009) Integrating pests and pathogens into the climate change/food security debate. J Exp Bot 60:2827–2838

Huang YJ, Evans N, Li ZQ, Eckert M, Chevre AM, Renard M et al (2006) Temperature and leaf wetness duration affect phenotypic expression of *Rlm6*-mediated resistance to *Leptosphaeria maculans* in *Brassica napus*. New Phytol 170:129–141

Hughes DJ, West JS, Atkins SD, Gladders P, Jeger MJ, Fitt BDL (2011) Effects of disease control by fungicides on Greenhouse Gas (GHG) emissions by UK arable crop production. Pest Manag Sci 67:1082–1092

Hulme M, Jenkins GJ, Lu X, Turnpenny JR, Mitchell TD, Jones RG et al (2002) Climate change scenarios for the United Kingdom: the UKCIP02 scientific report. Tyndall Centre for Climate Change Research, School of Environmental Sciences, University of East Anglia, Norwich, p 120

Jackson J, Li Y, Passant N, Thistlethwaite G, Thomson A, Cardenas L (2007) Greenhouse gas inventories for England, Scotland, Wales and Northern Ireland: 1990–2005. AEA Environment and Technology, Didcot, p 40, http://nora.nerc.ac.uk/2230/

Jamieson P, Semenov MA, Brooking I, Francis G (1998) SIRIUS: a mechanistic model of wheat response to environmental variation. Eur J Agron 8:161–179

Jones R, Noguer M, Hassell DC, Hudson D, Wilson SS, Jenkins GJ et al (2004) Generating high resolution climate change scenarios using PRECIS. Met Office Hadley Centre, Exeter, p 40

Lin BB, Chappell MJ, Vandermeer J, Smith G, Quintero E, Bezner-Kerr R et al (2011) Effects of industrial agriculture on climate change and the mitigation potential of small-scale agroecological farms. CAB Rev Perspect Agr Vet Sci Nutr Nat Resour 6:1–18

Madgwick JW, West JS, White RP, Semenov MA, Townsend JA, Turner JA et al (2011) Impacts of climate change on wheat anthesis and fusarium ear blight in the UK. Eur J Plant Pathol 130:117–131

Mahmuti M, West JS, Watts J, Gladders P, Fitt BDL (2009) Controlling crop disease contributes to both food security and climate change mitigation. Int J Agr Sustain 7:189–202

Nakicenovic N, Alcamo J, Davis G, Vries B, de Fen-hann J, Gaffin S et al (2000) Emissions scenarios. In: Nakicenovic N, Swart R (eds) Special report of the Intergovernmental Panel on Climate Change. Cambridge University Press, Cambridge

Oerke EC (2006) Crop losses to pests. J Agric Sci 144:31–43

Schmidhuber J, Tubiello FN (2007) Global food security under climate change. Proc Natl Acad Sci U S A 104:19703–19708

Semenov MA (2007) Development of high-resolutionUKCIP02-based climate change scenarios in the UK. Agr Forest Meteorol 144:127–138

Smith P, Martino D, Cai Z, Gwary D, Janzen H, Kumar P et al (2008) Greenhouse gas mitigation in agriculture. Philos Trans R Soc B 363:789–813

Stern N (2007) The economics of climate change: the Stern review. Cambridge University Press, Cambridge

Strange RN, Scott PR (2005) Plant disease: a threat to global food security. Annu Rev Phytopathol 43:83–116

West JS, Holdgate S, Townsend JA, Edwards SG, Jennings P, Fitt BDL (2012a) Impacts of changing climate and agronomic factors on fusarium ear blight of wheat in the UK. Fungal Ecol 5:53–61

West JS, Townsend JA, Stevens M, Fitt BDL (2012b) Comparative biology of different plant pathogens to estimate effects of climate change on crop diseases in Europe. Eur J Plant Pathol 133:315–331

Xu XM, Nicholson P (2009) Community ecology of fungal pathogens causing wheat head blight. Annu Rev Phytopathol 47:83–103

Xu XM, Monger W, Ritieni A, Nicholson P (2007) Effect of temperature and duration of wetness during initial infection periods on disease development, fungal biomass and mycotoxin concentrations on wheat inoculated with single, or combinations of Fusarium species. Plant Pathol 56:943–956

Zhang X, Halder J, White RP, Hughes DJ, Ye Z, Wang C et al (2014) Climate change increases risk of fusarium ear blight on wheat in central China. Ann Appl Biol 164:384–395

Transcriptomics and Genetics Associated with Plant Responses to Elevated CO_2 Atmospheric Concentrations

Amanda P. De Souza, Bruna C. Arenque, Eveline Q.P. Tavares, and Marcos S. Buckeridge

Introduction

The atmospheric CO_2 levels are predicted to increase during the coming years, altering temperature and precipitation patterns (Solomon et al. 2007). Because plants are very responsive to environmental changes, it is not surprising that these three factors—CO_2, water and temperature—will affect many aspects of the plants, including their physiological responses and their gene expression patterns.

At the same time, the steady rise in the world population presses for increasing demands for food, feed, fiber, and (bio) energy, all of them associated with the need of sustainable options to maintain the natural resources. This exerts, and will do so even stronger in the future, a high pressure on global agricultural productivity. According to Gruskin (2012), humankind will need a 70 % increase in food production to cope with the expected increases in population up to 2040. However, the present perspectives are that current crops are not well adapted to face the expected changes in atmospheric [CO_2] within the next decades (Leakey and Lau 2012). These predictions highlight the importance and the urgency to enhance traditional breeding and biotechnology techniques capable of improving adaptation of crops to the climate variables in order to increase food, feed, fiber, and energy security in the coming decades.

Changes in atmospheric CO_2 concentration ([CO_2]) have direct and immediate effects on plants, but their responses can be modified by feedback mechanisms depending on timescale (Medlyn and McMurtrie 2005). Therefore, different responses can be associated with functional types, growth forms, species, cultivars and genotypes (Wang et al. 2012). In experiments under elevated [CO_2],

A.P. De Souza, Ph.D. • B.C. Arenque, Ph.D. • E.Q.P. Tavares, Ph.D.
M.S. Buckeridge, Ph.D. (✉)
Botany Department, University of São Paulo,
Rua do Matão, 277, Butantã, Cidade Universitária, Sao Paulo 05508-090, Brazil
e-mail: msbuck@usp.br

D. Edwards, J. Batley (eds.), *Plant Genomics and Climate Change*,
DOI 10.1007/978-1-4939-3536-9_4

physiological parameters have been intensely explored, while there is a large gap in research including transcriptomics and genetics information. In order to use the biotechnological techniques necessary to improve modern crops, it is crucial to have access to the gene expression and genetics information, since there is still missing data concerning both aspects, which generally precludes the prediction of how plants will cope with environmental changes (Ahuja et al. 2010).

In this chapter, we describe some of the main physiological responses observed in elevated [CO_2] and discuss how the transcriptomics data and genetics are associated with those responses, highlighting the aspects that they could help in the improvement and adaptation of agricultural systems that are facing the increase in atmospheric [CO_2]. We expect that this could help to guide measures related to research improvement in agricultural biotechnology in a scenario of rapid changes that the world has been undertaking during the end of the twentieth and the twenty-first Century.

Main Physiological Responses of Plants Cultivated in Elevated [CO_2]

One of the well-characterized physiological effects of elevated [CO_2] in plants is related with the photosynthesis process. In C_3 plants, the assimilation of CO_2 usually increases, driven by a reduction of the use of CO_2 due to photorespiration. Photorespiration occurs due the oxygenation reaction that Rubisco (ribulose 1,5 bisphosphate oxygenase/carboxylase) catalyzes. When oxygen is assimilated, instead of two molecules of 3-phosphoglycerate, Rubisco leads to the production of RuBP 3-phosphoglycerate and 2-phosphoglycolate, reducing net CO_2 assimilation (Stitt 1991). Once the atmospheric [CO_2] increases, the competition between O_2 (ca. 21 % of the atmosphere) and CO_2 (ca. 0.04 % of the atmosphere) for the active site of the enzyme is reduced, decreasing photorespiration (Andrews and Lorimer 1978, 1987) and leading to an increase in CO_2 assimilation of 33 %, on average (Wand et al. 1999).

While in C_3 plants the photosynthetic response under elevated [CO_2] is well known, in C_4 plants there is no consensus about the nature of those responses (Leakey 2009). In C_4 plants, the primary enzyme that captures the CO_2 molecule is phosphoenolpyruvate carboxylase (PEPc), which has higher affinity for this substrate and at the same time, in PEPc, O_2 does not compete for the active site of the enzyme. C_4 plants also have a mechanism that concentrates CO_2 around Rubisco in specialized bundle sheath cells, increasing the [CO_2] more than five times, which strongly inhibits photorespiration (Furbank and Hatch 1987). For these two reasons, it is believed that C_4 plants cannot present any response related with photosynthesis process to elevated [CO_2]. However, stimulation in C_4 photosynthesis has been observed in various experiments described in literature (Wand et al. 1999; Watling et al. 2000; Vu et al. 2006; De Souza et al. 2008), leading to an average improvement of 25 % in plant biomass (Wand et al. 1999). The stimulation is attributed to a range of factors including: (a) direct stimulation (i.e. via Rubisco) due to the operating intracellular [CO_2] (c_i) being not saturated under ambient [CO_2] (Watling and Press 1997; Ziska and Bunce 1997; Moore et al. 1998); (b) altered bundle sheath leakiness (Watling et al. 2000), i.e. the high pressure in these cells leads CO_2 to escape to mesophyll so that it cannot be captured directly by Rubisco; (c) C_3-like photosynthesis in young C_4 leaves (Ziska et al. 1999), i.e. young leaves would not have anatomy prepared to perform C_4 system efficiently; (d) improvement in performance of electron transport chain (De Souza et al. 2008); and (e) indirect effects by improvement in water relations due the reduction of stomatal conductance (Ghannoum et al. 2000; Leakey et al. 2006).

In both C_3 and C_4 plants, the reduction in stomatal conductance is one of the main effects observed when plants are cultivated under elevated [CO_2] (Ainsworth and Long 2005). According to Wong et al. (1979), stomata tend to respond to c_i in such a way as to maintain a constant c_i:c_s (intracellular [CO_2]: leaf surface [CO_2]) ratio. When plants are cultivated under elevated [CO_2] and it is higher at the leaf surface (i.e. a lower c_i:c_s), a series of physiological mechanisms take place, involving hormones, which regulate the stomatal aperture and culminates with the reduction of CO_2 entrance in the cells (Field et al. 1995). In the long-term, the decrease in stomatal conductance can also be associated with a reduction in stomatal density and/or stomatal number (Drake et al. 1997). A consequence of the reduction in stomatal conductance is the decrease in water loss by transpiration that leads to an improvement in water use efficiency (e.g. carbon assimilation per volume of water transpired). In general, the improvement in water use efficiency in elevated [CO_2] is thought to alleviate the negative responses caused by drought, therefore maintaining growth and productivity under water stress (Leakey et al. 2006; Ainsworth and Long 2005; Wall et al. 2011).

The increase in CO_2 uptake by photosynthetic assimilation or by reduction in carbon losses during drought stress observed in elevated [CO_2], in most cases lead to an increment in plant growth. The additional carbon is allocated into leaves, stem and/or roots, promoting an increase in total biomass of 49 % (Poorter and Pèrez-Soba 2002). This growth stimulation can be associated with increase in leaf area, leaf number, leaf thickness, and stem, branch or root elongation, which are directly related with processes like cell division, cell expansion and cell differentiation (Pritchard et al. 1999).

The surplus of carbon observed in plants cultivated under elevated [CO_2] can also be stored as non-structural carbohydrates—glucose, fructose, sucrose and starch. In experiments in which long-term exposure of plants to elevated [CO_2] were followed, the increase in non-structural carbohydrates led to acclimation, a process that includes: (a) decrease in proteins, and nitrogen contents in leaves, (b) reduction of photosynthetic rate, (c) increase or decrease in respiration, (d) decrease or cessation of plant growth (Stitt 1991; Poorter and Pèrez-Soba 2002; Gifford et al. 2000; Moore et al. 1999). The rate of acclimation is closely related to the capacity of plant sinks to store carbohydrates and/or to establish new sinks (Stitt 1991).

The reduction in photosynthesis by the acclimation process is associated with different mechanisms (Araya et al. 2006). One of them is related to the inhibition of sucrose synthase activity by high carbohydrate concentration. Without the activity of this enzyme, there is an accumulation of phosphate sugars in the cytosol that reduces the availability of inorganic phosphate (P_i) to metabolism. As the P_i is directly related to the regeneration of ribulose 1,5 bisphosphate (RuBP), which is the substrate for Rubisco, the decrease in availability of this molecule ends up decreasing Rubisco activity (Stitt 1986). Alternatively, it is also known that the accumulation of hexoses can reduce photosynthesis rates by repressing expression of genes related to the photosynthetic process via hexokinase (Cheng et al. 1998). This mechanism is called, in general, *sugar sensing*.

The increase in non-structural carbohydrates in plants cultivated in elevated [CO_2] can also increase the respiratory substrate availability and therefore reduces respiratory capacity (Griffin et al. 2001). At the same time, leaf nitrogen status and protein contents are often lower, reducing the demand for respiratory ATP and therefore decreasing respiration (Norby et al. 1999; Ellsworth et al. 2004). Thus, the balance of these responses is what determines the alteration in respiration rates under elevated [CO_2].

The higher accumulation of carbon at elevated [CO_2] dilutes the plant nutrients concentrations, artificially increasing the efficiency of the plant to use these nutrients (Norby et al. 2005; Seiler et al. 2009). However, in the long-term responses, the nutrient dilution by elevation of [CO_2] can be excessive and could restrict the carbon uptake (i.e. increasing the speed of acclimation process) (Luo et al. 2004). Nitrogen concentration and allocation have been extensively evaluated in elevated [CO_2]. Data from a meta-analysis using 75 published studies showed that nitrogen content under elevated [CO_2] reduced independently of the plant type, which represents an increase in nitrogen use efficiency (Cotrufo et al. 1998). Due to the fact that the movement of nutrients from soil to plant root is dependent of mass flow (Kabata-Pendias and Pendias 2001), nutrient uptake may also be reduced due the decreased leaf transpiration rates in elevated [CO_2]. Although other nutrients, such as phosphorus, calcium, boron, copper, iron, potassium, magnesium, manganese, sulphur and zinc have received little attention, it is known that their dilution in the plant at elevated [CO_2] varies according the functional groups and among organs (Duval et al. 2012).

After the publication of thousands of papers describing experiments in which physiological responses under elevated [CO_2] were the main focus, it seems to be reasonably clear that most plants respond to elevated [CO_2] in a very similar fashion. The general pathways of the *fertilization responses* as well as the *acclimation responses* are pictured in Fig. 1 along with the related genes that will be discussed below. In the next section, in which we examine reported mechanisms and the general patterns of gene expression that have been demonstrated to be associated with the main plant models and crops used for research. Gene expression is put forward under the light of the physiological mechanisms described in the present section.

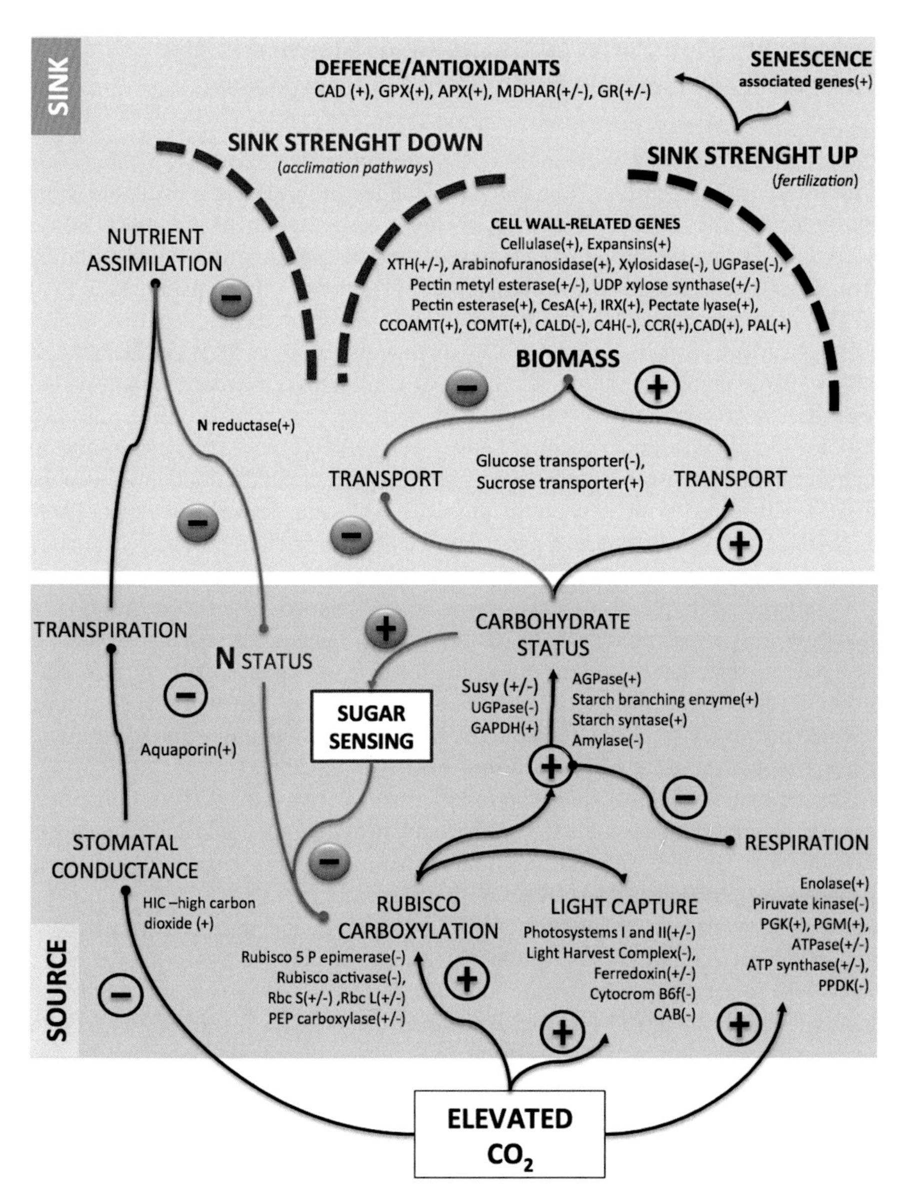

Fig. 1 Schematic representation of the compiled responses of plants to elevated [CO_2]. Physiological responses are linked through pathways related to CO_2 fertilization (*black*) and acclimation (*red*). The genes thought to be related to every physiological response are listed below each response. Upregulation (+) or downregulation (–) represent findings in the literature that have been reported for a given species (see Table 1 for references). The idea of source-sink relationship shown in this figure is a theoretical elaboration on the basis of mostly data obtained from plant leaves. We artificially separate leaf-related processes (photosynthesis and transpiration), i.e. source features, from more sink related processes such as cell wall metabolism and nutrient acquisition. Processes that probably occur in the entire plant, such as respiration, defense and senescence are left where better connections could be made for the sake of understanding the associations of the pathways of response to [CO_2]. The *sugar sensing* mechanism is put forward as a central regulator, along with an "N sensing" system, that seem to be key in acclimation processes in plants

Gene Expression Patterns Changing at Elevated [CO_2]: How Are They Coupled to Physiological Responses?

Physiological responses at elevated [CO_2] are generally followed by changes in plant transcriptional profile. Nonetheless, much less information is available about the molecular mechanisms that mediate those responses in physiological scale. According to data published so far, gene expression profiling can be obtained by several methods (Busch and Lohmann 2007) among which quantitative PCR (qPCR) and microarrays are, to the present date, the main techniques used to analyze gene expression in elevated [CO_2]. Although real time PCR can be used to screen a large number of genes (Czechowski et al. 2004), microarray analysis has been the most common method used to perform transcript profiling of plant responses to global climate change (Leakey et al. 2009a). With the development of high-throughput next-generation sequencing techniques and the continuous decline in sequencing costs with a concomitant increase of sequence reads lengths (Thudi et al. 2012), in the near future, a great increase in transcription profiling evaluation is expected in a higher number of environmental conditions.

Compared with physiological responses, modulation of gene expression is thought to respond more rapidly to environmental changes. As soon as the plant is changed to a different environmental condition (e.g. elevated [CO_2]), the new condition is sensed, initiating a signal transduction pathway that will drive, in most cases, to metabolic adjustments that culminate in plant adaptation, which represents a new steady-state in the new conditions (Leakey et al. 2009a).

The evaluation of gene expression in experiments at elevated [CO_2] is commonly focused on long-term responses, aimed at understanding the molecular controls of physiological effects. Considering that experiments that evaluated long-term responses usually detect fewer genes with differential expression than experiments that evaluate short-term responses (Campos et al. 2004), and that microarray hybridization analyses do not always contemplate the entire genome/transcriptome of the plants tested, generally a rather low number of genes is observed to change their expression at elevated [CO_2]. In maize, for instance, only 5.2 % (387 of 7442) of the genes that were analyzed underwent changes in transcriptional pattern at elevated [CO_2] (Kim et al. 2006). Only a total of 33 from 3598 genes (~1 %) evaluated by microarray in sugarcane was found as differentially expressed under elevated [CO_2], although the genes present in the microarray used in this experiment represented less than 10 % of the sugarcane transcriptome (De Souza et al. 2008). In poplar, Taylor et al. (2005) found approximately 0.06 % of the genes were sensitive to high [CO_2] in young leaves and around 0.1 % in semi-mature leaves. Furthermore, the largest fold changes in transcript abundance due the CO_2 treatment are typically twofold (Leakey et al. 2009a).

Regardless of the small number of genes and the low amplitude of fold change observed at elevated [CO_2] experiments, a range of information has been explored and associated with physiological responses. Table 1 summarizes the expression profile evaluated by microarray approaches related to the main physiological

Table 1 Expression profile of microarray data obtained from experiments in elevated [CO_2]

Category	Plant	Expression profile	Reference
Model system	*Arabidopsis thaliana*	*Upregulation*: SuSy, glucose transporter, PGM, Nitrate reductase, nitrite reductase, glutamate synthase, glutamine synthase *Downregulation:* UXS, UGPase, Photosystem I, CAB, LHC, Fd, ATP synthase, amylase, rbcL, rbcS	Li et al. (2007)
Horticulture	*Betula pendula*	*Upregulation*: Photosystem II *Downregulation*: Xylosidase, amylase, photosystem I, senescence-associated genes, NADH dehydrogenase, cytocrome oxidase, NADH plastoquinone oxidoreductase	Kontunen-Soppela et al. (2010)
Food	*Oryza sativa*	*Upregulation*: Photosystem II, sucrose transporter, rbcS, rbcL, starch synthase *Downregulation*: CAB, nitrate transporter, rbcS, photosystem II	Fukayama et al. (2011)
Food; Bioenergy; Feed	*Glycine max*	*Upregulation*: ATP synthase, AGPase, starch synthase, starch branching enzyme, phosphoglycerate mutase, PEP carboxylase, SuSy, phosphoglycerate kinase, phosphoglucomutase, UDP-glucose pyrophosphorylase, UTP glucose-1-phosphate uridyltransferase, phosphofructokinase, pyruvate kinase, NADP malic enzyme, electron carrier-related genes, glucose-6-phosphate translocator, sucrose phosphatase, enolase, malate dehydrogenase, GPX7ch, APX4ch, MDHARpx, GRch *Downregulation*: CAB, cytocrome b6f, photosystem I and II, rbcL, PPDK, phosphoglucomutase, MDHARcyt, GRcyt	Ainsworth et al. (2006); Leakey et al. (2009b)
Food; Bioenergy; Feed	*Zea mays*	*Upregulation*: SuSy3, rbcL, rbcS *Downregulation*: GAPDH, photosystem II, rbcS, vacuolar H^+ pyrophosphatase, CAB, XTH	Kim et al. (2006)
Food; Bioenergy; Feed	*Saccharum* sp.	*Upregulation*: Fd, PEP carboxylase, Photosystem I and II, arabinofuranosidase, invertase, XTH *Downregulation*: Glucose-6-phosphate dehydrogenase	De Souza et al. (2008)
Bioenergy	*Populus deltoides*	*Upregulation*: XTH, cellulase, COMT, CAD, CCoAOMT	Druart et al. (2006)
Bioenergy	*Populus* × *euramericana*	*Upregulation*: GAPDH, PAL, starch synthase, CAD, PME, UXS, CCR *Downregulation*: Pyruvate kinase, potassium transporter, photosystem II and II, rbcS, PEP-CK, SuSy, RCA	Taylor et al. (2005); Tallis et al. (2010)
Bioenergy	*Populus tremuloides*	*Upregulation*: Expansin, pectin esterase, CesA, UTP-glucose-1-phosphate-uridyltransferase, IRX, pectate lyase, SuSy, XTH, aquaporin, enolase *Downregulation*: Ribulose-5-phosphate epimerase, rbcS, photosystem I and II, rbcL, CAB, PME, ATPase, aminomethyltransferase	Gupta et al. (2005); Cseke et al. (2009); Wei et al. (2013)

Only genes related with physiological parameters described in the text were cited in this table

SuSy—sucrose synthase; PGM—phosphoglycerate mutase; UXS—UDP xylose synthase; UGPase—UGP-glucose pyrophosphorylase; CAB—*Chlorophyll* a/b *binding*; LHC—light harvesting complex; Fd—ferredoxin; rbcL—rubisco large subunit; rbcS—rubisco small subunit; AGPase—ADP-glucose pyrophosphorylase; GPX—glutatione peroxidase; APX—ascorbate peroxidase; MDHAR—monodehydroascorbate reductase; GR—glutatione reductase; PPDK—PEP *phosphate dikinase*; GAPDH—glyceraldehyde 3-phosphate dehydrogenase; XTH—xyloglucan transglycosilase/hydrolase; CoMT—*caffeate o-methyl transferase*; CAD—*cinnamyl alcohol dehydrogenase*; CCoAOMT—*caffeoyl-CoA-O-methyltransferase*; PAL—*phenilalanine ammonia lyase*; PME—*pectin methyl esterase*; CCR cynnamoyl-CoA reductase; PEP-CK—phosphoenol pyruvate carboxykinase; RCA—*Rubisco* activase; CesA—celulose synthase A; IRX—arabinoxylan biosynthesis gene; ch—chroloplast isoform; px—peroxisome isoform; mt—mitochondria isoform processes found to change under elevated [CO_2]. Species that have already had their gene expression profile at high [CO_2] evaluated are used for different purposes such as food, feed, fiber, bioenergy and horticulture. Differentially expressed genes between ambient and elevated [CO_2] are related to photosynthesis (e.g. light capture and carboxylation related), non-structural carbohydrates (e.g. starch and sucrose related), structural carbohydrates (e.g. cell wall related), energetic metabolism (e.g. mitochondria related), transport, defense and antioxidants (e.g. secondary metabolism related), nutrient acquisition and senescence categories. In most of cases, they seem to follow the same pattern of changes that are associated with the physiological effects described in the previous section (Fig. 1). For example, in sugarcane De Souza et al. (2008) found a higher expression of genes related with photosynthesis and carbohydrate metabolism in elevated [CO_2] at the same time as photosynthesis was higher in this treatment. When a significant difference in biomass and height between treatments (elevated vs. ambient [CO_2]) was observed, the main categories affected were cell cycle, development, protein and nucleic acid metabolisms. Interestingly, these authors found that chlorophyll *a-b*-binding protein, ferredoxin, photosystem I reaction center subunit N and photosystem II subunit K, all of these genes related to electron transport system, were up-regulated in leaves of sugarcane under high [CO_2] at the same time that the electron transport rate was higher in this treatment (De Souza 2011).

One mechanism that is well characterized at the molecular level and also shows correlation with the physiological responses observed in plants grown under elevated [CO_2] is the acclimation process of photosynthesis. Carbohydrates mediate the regulation of a series of photosynthetic and non-photosynthetic genes (Koch 1996), so that under elevated [CO_2] the carbohydrate production is thought to be the main factor that regulates the acclimation process (i.e. the *sugar sensing* mechanism). After a stimulation of photosynthesis, growth and carbohydrate production in plants exposed to high [CO_2] (Fig. 1), a decrease in the levels of mRNAs encoding for certain photosynthetic genes is observed. Rubisco activase and Rubisco 5-phosphate epimerase are usually decreased, while Rubisco small subunit (RbcS) and Rubisco large subunit (RbcL) are affected in a number of ways (Moore et al. 1998). *RbcS* and *RbcL* transcript levels can remain unchanged or display an increase of one of the subunits without changing another or increase one of the subunits and at the same time decrease another. This suggests the involvement of multiple controlling events in the photosynthetic responses at the molecular level. Genes involved with other components of photosynthetic apparatus can also have their expression reduced during the acclimation process (e.g. chlorophyll a-b binding (CAB), light harvest complex, cytochrome B6f).

Although many sugar sensors have been identified in plants, under elevated [CO_2] the cascade of signaling mediated by carbohydrates seems to be initiated by hexokinases (Cheng et al. 1998; Moore et al. 2003). If the sucrose cycling is increased (what usually occurs under elevated [CO_2]) and there is little demand for carbon sink organs, the relatively high level of hexoses in the cytosol of source cells leads to an increase in hexokinase activity that will result in a repression of photosynthetic genes. On the other hand, if the sink demand is significant, the hexokinase signaling is reduced and, consequently, the photosynthetic acclimation process is delayed.

Although the correlation between gene expression and physiological parameters appears to be, in a first glance, straightforward, it is important to emphasize that the evaluation of those responses represent only a snapshot of the phenomenon. The interpretation of transcript abundance should consider the post-transcriptional and post-translational regulation as well as the transcript half life and turnover rate of the encoded proteins and metabolites involved in such pathways, which can regulate the expression of the given gene (Leakey et al. 2009a). The interpretation should also consider that the circadian clock controls many genes and they will not necessarily present the same pattern of response during all hours of the day (Harmer et al. 2000; Michael and McClung 2003). This is especially true for leaves, where the clock ticking is much more apparent than in stem and roots (De Souza 2011). Furthermore, transcription data is strongly biased towards leaves and as a result, the responses of gene expression in sink organs such as stem or roots are still unknown. As it is clear that the sink organs play an essential role during the regulation processes under elevated [CO_2] the lack of transcription profile data of sink-related genes poses a serious limitation to the interpretation of acclimation effects via *sugar sensing*. Thus, knowing which and how genes are being expressed in sink organs is fundamental to elucidate the molecular coordination that occurs under elevated [CO_2].

Another limitation in gene expression responses is the lack of information about gene function. Due the high number of *unknown* genes, many mechanisms that are changing under a different CO_2 concentration cannot be identified. In maize, for example, Kim et al. (2006) found 387 genes differentially expressed at elevated [CO_2], but 73 % of these genes lacked annotation. Thus, genome sequencing projects and manual annotation efforts are essential to understand the changes that occur in the transcriptional scale in a changing environment.

In addition to gene expression modulation in elevated [CO_2], genetic variability could affect the response of plants in this condition. In the next section, we will discuss these findings and also consider the effect that the stress caused by CO_2 can have on plant genetics via epigenetic events.

How Can Genetic Variability Affect the Responses of Plants Under Elevated [CO_2], and How Can Elevated [CO_2] Lead to Genetic Changes?

After the two previous sections, there is little doubt that plants grown in elevated [CO_2] will have their physiology and gene expression altered. However, a question still remains: what is the genetic basis that controls those responses?

According to Richardson et al. (2012), three biological processes can be involved in plant genetic responses under climate changes: (a) phenotypic plasticity, (b) dispersal of propagules (e.g. change gene flow through seed and/or pollen), and (c) genetic changes (e.g. gene mutation), that could act separately or potentially interact with each other. In the same way as for physiological responses, the genetic mechanisms that will act to cope with the environmental changes are likely to vary depending on the time scale considered. Among these processes, the phenotypic plasticity, which is defined as the capacity of a particular genotype to produce varied phenotypes in response to an environmental change (Pigliucci 2001), is thought to be a crucial determinant of plant responses in the short-term (Nicotra et al. 2010). Also, Matesanz et al. (2010) argue about the possibility that climate changes can influence the evolution of phenotypic plasticity, since it could have a benefit of the alleles that confer such plasticity.

Phenotypic plasticity can be temporary (non-heritable) or heritable, the latter being possible through some epigenetic events (Richardson et al. 2012). Epigenetic modifications are based on a set of processes that can activate, reduce or completely disable the activity of particular genes (Bossdorf et al. 2008). These processes involve: (a) methylation of cytosine residues in the DNA molecule, (b) acetylation or methylation of histone proteins, (c) regulatory mechanisms mediated by small RNA molecules, and (d) transposable element activation. Although it is known that epigenetic changes can act in response to climate change (e.g. temperature) (Nicotra et al. 2010), these mechanisms are poorly understood under elevated [CO_2]. May et al. (2013) identified microRNA levels changing at elevated [CO_2] in *Arabidopsis thaliana*. Among those, miR156/157 and miR172 families were strongly correlated with changes in transcript levels of their target genes, possibly acting as regulators of early flowering induced by rising [CO_2].

Genotype variation is also an important factor that influences the short-term responses at elevated [CO_2], since different genotypes can display distinct responses under those conditions (Springer and Ward 2007; Li et al. 2006, 2008). In *Arabidopsis thaliana*, Li et al. (2006) showed that three genotypes (Wassilewskija [WS], Columbia-0 [Col-0], and Cape Verde Island [Cvi-0]) presented different responses in transcriptional and metabolic profiles at elevated [CO_2]. For example, WS and Col-0 showed an increase in non-structural carbohydrates and amino acids whereas Cvi-0 showed a consistent and significant decrease in these compounds under elevated [CO_2], flowering later than the other two genotypes. According to the authors, these different strategies of response might reflect the environmental condition that the genotypes are adapted to (i.e. WS and Col-0 are weedy species that recycle nutrients and grow faster, and Cvi-0 is adapted to grow in dry and high light conditions).

Alternatively, some species may not present distinctive responses among genotypes. Bourgault et al. (2013) evaluated 20 wheat lines at elevated [CO_2] that have contrasting tillering propensity, soluble carbohydrate accumulation in the stem, early vigor, and transpiration efficiency. They found no significant responses to elevated [CO_2] for any of the analyzed lines, suggesting that in wheat the genotype will not interfere in biomass production and yield. Thus, the variation of responses among genotypes seems to be related with other aspects that are species-dependent.

One of the methods to identify genome regions that can be associated with genotype-dependent responses is the quantitative trait loci (QTL) mapping. According to Franks and Hoffmann (2012) QTL identification as well as artificial selection and experimental evolution, gene polymorphisms or DNA sequence comparisons and associating mapping are the most used methods to evaluate genetic responses to climate changes in manipulated populations, as is the case of crops. Specifically in experiments where plants are grown at elevated [CO_2], QTL mapping has been used to identify genome regions related with stomatal and epidermal cell development (Ferris et al. 2002), leaf growth and senescence (Rae et al. 2006), biomass increment (Rae et al. 2007), yield (Fan et al. 2008) and flowering (Ward et al. 2012). The understanding of genome regions that are involved with important physiological traits at elevated [CO_2] is relevant to plant productivity as well as to characteristics that determine plant survival, fitness and interaction with pests and pathogens (Ward and Kelly 2004).

Ward et al. (2012) analyzed 179 recombinant inbred lines of *Arabidopsis thaliana* and mapped only one QTL related with changes in flowering time under elevated [CO_2]. Due to the fact that only one significant QTL has been identified, the authors suggested that the change in flowering time at elevated [CO_2] is likely the result of polymorphism in a single one or from a small number of genes. In fact, they conducted analysis of knockout mutants, focusing on five flowering genes associated with the identified QTL and observed significant differences only at MOTHER OF FT AND TFL1 (MFT) mutant, which flowered much earlier in high [CO_2], suggesting that this gene could be a candidate for altered flowering time at elevated [CO_2].

In *Populus*, 238 genotypes from an F2 pedigree (Family 331—*P. trichonocarpa* × *P. deltoides*) were evaluated under elevated [CO_2] and a series of putative QTL to adaxial and abaxial stomatal control was identified (Ferris et al. 2002). These QTL are positioned close to each other, suggesting that the same part of the genome can control the stomatal initiation on both sides of leaves. Also, they found that epidermal cell area, cell number and leaf area are co-located in the *Populus* genome, mapping to almost the same position, suggesting that just one QTL affects all these leaf traits. Later on, Rae et al. (2006) identified, in the same group of poplar genotypes, 28 responsive QTL for 19 traits, including stomatal index, stomatal density, leaf area, leaf width, leaf expansion, leaf cell number and senescence. However, from these 28 responsive QTL, only 21 were found to be co-located across ambient and elevated [CO_2]. For example, a QTL related to stomatal traits (i.e. stomatal index and stomatal density), was mapped at ambient [CO_2] but the authors found no evidence of this QTL being relevant at elevated [CO_2], suggesting that stomatal index and stomatal density is differentially controlled across both CO_2 concentrations. They also identified six QTL related to senescence and showed that a QTL responsive to elevated [CO_2]

is located in a distinct group in relation to ambient [CO_2], indicating that the region which controls the leaf senescence at elevated [CO_2] is not the same region observed in ambient [CO_2]. This suggests that the increase in atmospheric [CO_2] could drive microevolutionary adaptations as already explored by other authors (Gienapp et al. 2008; Pauls et al. 2013). In addition, Rae et al. (2007) identified a QTL related to stem diameter of a specific group in ambient [CO_2] but not in elevated [CO_2], implying that this region of the genome could not be expressed under high [CO_2].

Although some publications have described connections between traits and genotypes or specific QTL and genes, large gaps remain in identifying the genetic basis of responses to different climatic conditions (Franks and Hoffmann 2012). In that case, the new experiments under elevated [CO_2] should consider the following points: (a) evaluation of inter and intraspecific variation to quantify genetic-dependent responses, (b) use the new available tools to compare physiological traits, metabolites, transcripts and genomes, and (c) evaluate the response to elevated [CO_2] along with water and temperature changes that are tightly connected to the climate change scenario (Ainsworth et al. 2008).

Dealing with Big Data: *Systems Biology* Approach to Cope with Future of Global Climate Changes

Although nowadays there is a strong asymmetry between physiological information compared to gene expression and genetics of plants under elevated [CO_2], the new technologies (i.e. next-generation sequencing) that are now available will change this scenario. Nevertheless, in order to produce solid information about how plants will cope with the climate changes in the future, all these incoming data will have to be integrated. In addition, the integration should consider not only the effects of CO_2, but also the effects related to temperature and water as well as their possible combinations since they are altogether environmental stressors related to climate changes. Furthermore, different species, varieties, genetically transformed plants in both controlled conditions and field trials will need to be analyzed. Thus, it is likely that the most important issue in the next few years will be data analysis and integration.

In this context, *system biology* approaches, which involve an enormous effort of bioinformatics to construct networks with the data, seems to be the more efficient and realistic way to integrate all these responses. In Fig. 2 we illustrate a possible strategy to use *system biology* to cope with climate changes.

We believe that the knowledge generated from these analyses will allow the establishment of solid strategies capable to cope with the climate changes in the future, contributing to keep crop production growth steady and perhaps cope with population growth during the twenty-first century.

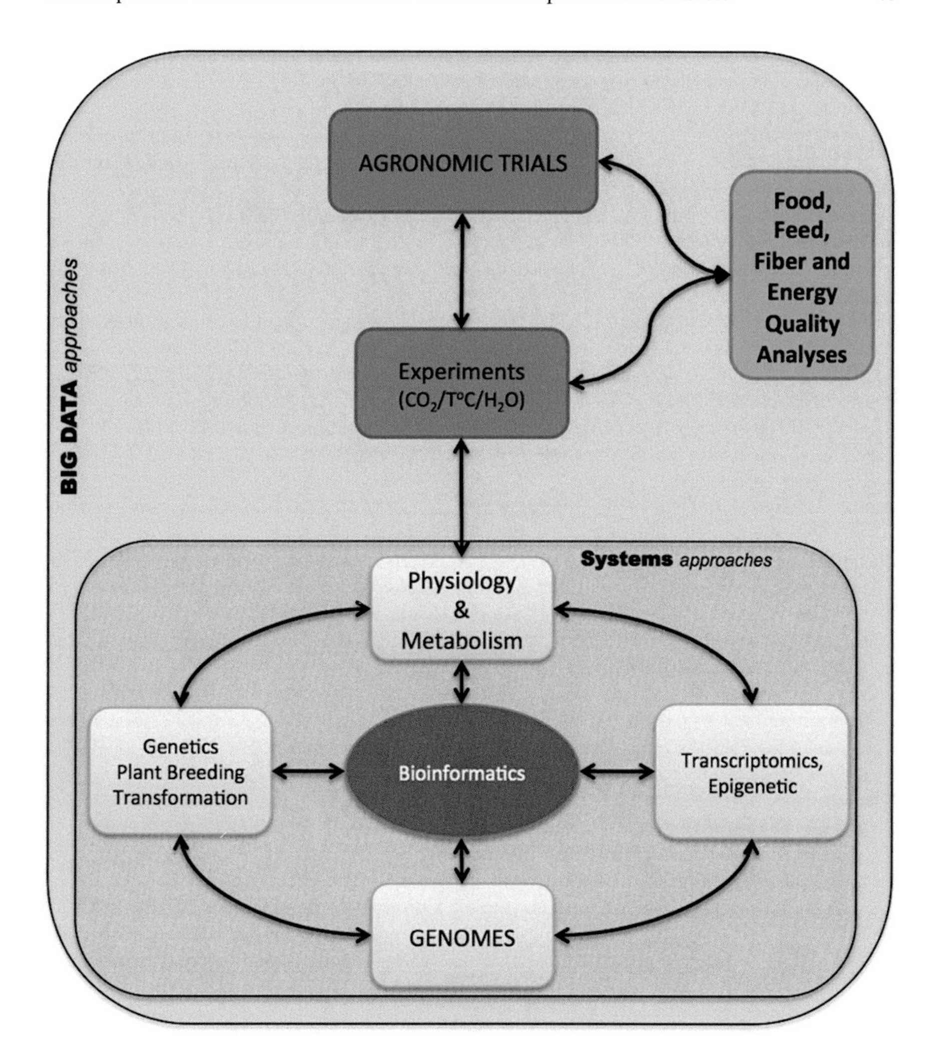

Fig. 2 Strategy for data integration and experimental approaches that could help to keep crop production in a scenario of Global Climate Changes. A *systems biology* approach integrating physiology, transcripts and genetics could serve as framework for experiments in elevated [CO_2] systems so that food, feed, fiber and bioenergy productivity and quality could be planed and controlled through an even larger integration of data considering other environmental stresses such as changes in temperature and water availability as well as other scales

References

Ahuja I, de Vos RCH, Bones AM, Hall RD (2010) Plant molecular stress responses face climate change. Trends Plant Sci 15(12):664–674

Ainsworth EA, Long SP (2005) What have we learned from 15 years of free air-CO_2 enrichment (FACE)? A meta-analytic review of the responses of photosynthesis, canopy properties and plant production to rising CO_2. New Phytol 165(2):351–372

Ainsworth EA, Rogers A, Vodkin LO, Walter A, Schurr U (2006) The effects of elevated CO_2 concentration on soybean gene expression. An analysis of growing and mature leaves. Plant Physiol 142(1):135–147
Ainsworth EA, Beier C, Calfapietra C, Ceulemans R et al (2008) Next generation of elevated [CO_2] experiments with crops: a critical investment for feeding the future world. Plant Cell Environ 31(9):1317
Andrews TJ, Lorimer GH (1978) Photorespiration still unavoidable? FEBS Lett 90:1–9
Andrews JT, Lorimer GH (1987) Rubisco: structure, mechanisms and prospects improvement. In: Haleh MD, Boardman NK (eds) Biochemistry of plants, vol 10. Academic, New York, NY, pp 132–207
Araya T, Noguchi K, Terashima I (2006) Effects of carbohydrate accumulation on photosynthesis differ between sink and source leaves of *Phaseolus vulgaris* L. Plant Cell Physiol 47:644–652
Bossdorf O, Richards CL, Pigliucci M (2008) Epigenetics for ecologists. Ecol Lett 11(2):106–115
Bourgault M, Dreccer MF, James AT, Chapman SC (2013) Genotype variability in the response to elevated CO_2 of wheat lines differing in adaptive traits. Funct Plant Biol 40(2):172–184
Busch W, Lohmann JU (2007) Profiling a plant: expression analysis in *Arabidopsis*. Curr Opin Plant Biol 10:136–141
Campos H, Cooper A, Habben JE, Edmeades GO, Schussler JR (2004) Improving drought tolerance in maize: a view from industry. Field Crop Res 90(1):19–34
Cheng SH, Moore BD, Seemann JR (1998) Effects of short- and long-term elevated CO_2 on the expression of ribulose-1,5-bisphosphate carboxylase/oxygenase genes and carbohydrate accumulation in leaves of *Arabidopsis thaliana* (L.) Heynh. Plant Physiol 116(2):715–723
Cotrufo MF, Ineson P, Scott Y (1998) Elevated CO_2 reduces the nitrogen concentration of plant tissues. Glob Change Biol 4(1):43–54
Cseke LJ, Tsai CJ, Rogers A, Nelsen MP, White HL, Karnosky DF, Podila GK (2009) Transcriptomic comparison in the leaves of two aspen genotypes having similar carbon assimilation rates but different partitioning patterns under elevated [CO_2]. New Phytol 182(4):891–911
Czechowski T, Bari RP, Stitt M, Scheible W-R, Udvardi MK (2004) Real-time RT-PCR profiling of over 1400 *Arabidopsis* transcription factors: unprecedented sensitivity reveals novel root- and shoot-specific genes. Plant J 38(2):366–379
De Souza AP (2011) Mecanismos fotossintéticos e relação fonte-dreno em cana de açúcar cultivada em atmosfera enrquecida em CO_2 [thesis]. Universidade de São Paulo, São Paulo (SP)
De Souza AP, Gaspar M, Silva EA, Ulian EC, Waclawovsky AJ, Nishiyama MY Jr et al (2008) Elevated CO_2 increases photosynthesis, biomass and productivity, and modifies gene expression in sugarcane. Plant Cell Environ 31(8):1116–1127
Drake BG, Gonzalez-Meler MA, Long SP (1997) More efficient plants: a consequence of rising atmospheric CO_2? Ann Rev Plant Physiol Plant Mol Biol 48:609–639
Druart N, Rodriguez-Buey M, Barron-Gafford G (2006) Molecular targets of elevated [CO_2] in leaves and stems of *Populus deltoides*: implications for future tree growth and carbon sequestration. Funct Plant Biol 33:121–131
Duval BD, Blankidhip JC, Dijkstra P, Hubgate BA (2012) CO_2 effects on plant nutrient concentration depend on plant functional group and available nitrogen: a meta-analysis. Plant Ecol 213(3):505–521
Ellsworth DS, Reich PB, Naumburg ES, Koch GW, Kubiske ME, Smith SD (2004) Photosynthesis, carboxylation and leaf nitrogen responses of 16 species to elevated pCO_2 across four free-air CO_2 enrichment experiments in forest, grassland and desert. Glob Change Biol 10:2121–2138
Fan G, Li X, Cai Q, Zhu J (2008) Detection of elevated CO_2 responsive QTLs for yield and its components in rice. Afr J Biotechnol 7(11):1707–1711
Ferris R, Long L, Bunn SM, Robinson KM, Bradshaw HD, Rae AM, Taylor G (2002) Leaf stomatal and epidermal cell development: identification of putative quantitative trait loci in relation to elevated carbon dioxide concentration in poplar. Tree Physiol 22(9):633–640
Field CB, Jackson RB, Mooney HA (1995) Stomatal responses to increased CO_2: implications from the plant to the global scale. Plant Cell Environ 18:1214–1225

Franks SJ, Hoffmann AA (2012) Genetics of climate change adaptation. Annu Rev Genet 46:185–208

Fukayama H, Sugino M, Fukuda T, Masumoto C, Taniguchi Y, Okada M, Sameshima R, Hatanaka T, Misoo S, Hasegawa T, Miyao M (2011) Gene expression profiling of rice grown in free air CO_2 enrichment (FACE) and elevated soil temperature. Field Crop Res 121(1):195–199

Furbank RT, Hatch MD (1987) Mechanism of C_4 photosynthesis – the size and composition of the inorganic carbon pool in bundle sheath cells. Plant Physiol 85(4):958–964

Ghannoum O, von Caemmerer S, Ziska LH, Conroy JP (2000) The growth response of C_4 plants to rising atmospheric CO2 partial pressure: a reassessment. Plant Cell Environ 23(9):931–942

Gienapp P, Teplitsky C, Alho JS, Mills JA, Merilä J (2008) Climate change and evolution: disentangling environmental and genetic responses. Mol Ecol 17(1):167–178

Gifford RM, Barrett DJ, Lutze JL (2000) The effects of elevated [CO_2] on the C:N and C:P mass ratios of plant tissues. Plant and Soil 224(1):1–14

Griffin KL, Anderson OR, Gastrich MD, Lewis JD, Lin GH, Schuster W et al (2001) Plant growth in elevated CO_2 alters mitochondrial number and chloroplast fine structure. Proc Natl Acad Sci U S A 98(5):2473–2478

Gruskin D (2012) Agbiotech 2.0. Nat Biotechnol 30(3):211–214

Gupta P, Duplessis S, White H, Karnosky DF, Martin F, Podila GK (2005) Gene expression patterns of trembling aspen trees following long-term exposure to interacting elevated CO_2 and tropospheric O_3. New Phytol 167(1):129–141

Harmer SL, Hogenesch JB, Straume M, Chang HS, Han B, Zhu T, Wang X, Kreps JA, Kay SA (2000) Orchestrated transcription of key pathways in Arabidopsis by the circadian clock. Science 290(5499):2110–2113

Kabata-Pendias A, Pendias H (2001) Trace elements in soils and plants. CRC Press, Boca Raton, FL

Kim SH, Sicher RC, Bae H, Gitz DC, Baker JT, Timlin DJ, Reddy VR (2006) Canopy photosynthesis, evapotranspiration, leaf nitrogen, and transcription profiles of maize in response to CO_2 enrichment. Glob Change Biol 12(3):588–600

Koch KE (1996) Carbohydrate-modulated gene expression in plants. Annu Rev Plant Physiol Plant Mol Biol 47:509–540

Kontunen-Soppela S, Riikonen J, Ruhanen H, Brosché M, Somervuo P, Peltonen P, Kangasjärvi J, Auvinen P, Paulin L, Keinänen M, Oksanen E, Vapaavuori E (2010) Differential gene expression in senescing leaves of two silver birch genotypes in response to elevated CO_2 and tropospheric ozone. Plant Cell Environ 33(6):1016–1028

Leakey ADB (2009) Rising atmospheric carbon dioxide concentration and the future of C_4 crops for food and fuel. Proc R Soc Biol Sci 276:2333–2343

Leakey ADB, Lau JA (2012) Evolutionary context for understanding and manipulating plant responses to past, present and future atmospheric [CO_2]. Phil Trans R Soc B 367(1588):613–629

Leakey ADB, Uribelarrea M, Ainsworth EA, Naidu SL, Rogers A, Ort DR, Long SP (2006) Photosynthesis, productivity and yield of maize are not affected by open-air elevation of CO_2 concentration in the absence of drought. Plant Physiol 140(2):779–790

Leakey ADB, Ainsworth EA, Bernard SM, Markelz RJC, Ort DR, Placella SA, Rogers A, Smith MD, Sudderth EA, Weston DJ, Wullschleger SD, Yuan S (2009a) Gene expression profiling: opening the black box of plant ecosystem responses to global change. Glob Change Biol 15(5):1201–1213

Leakey ADB, Xu F, Gillespie KM, MGrath JM, Ainsworth EA, Ort DR (2009b) Genomic basis for stimulated respiration by plants growing under elevated carbon dioxide. Proc Natl Acad Sci U S A 106(9):3597–3602

Li P, Sioson A, Mane SP, Ulanov A, Grothaus G, Heath LS, Murali TM, Bohnert HJ, Grene R (2006) Response diversity of *Arabidopsis thaliana* ecotypes in elevated [CO_2] in the field. Plant Mol Biol 62(4-5):593–609

Li P, Bohnert HJ, Grene R (2007) All about FACE-plants in a high-[CO_2] world. Trends Plant Sci 12(3):87–89

Li P, Ainsworth EA, Leakey AD, Ulanov A, Lozovaya V, Ort DR, Bohnert HJ (2008) Arabidopsis transcript and metabolite profiles: ecotype-specific responses to open-air elevated [CO_2]. Plant Cell Environ 31(11):1673–16787
Luo Y, Su B, Currie WS, Dukes JS, Finzi A, Hartwig U et al (2004) Progressive nitrogen limitation of ecosystem responses to rising atmospheric carbon dioxide. BioScience 54(8):731–739
Matesanz S, Gianoli E, Valladares F (2010) Global change and the evolution of phenotypic plasticity in plants. Ann N Y Acad Sci 1206:35–55
May P, Liao W, Wu Y, Shuai B, McCombie WR, Zhang MQ, Liu QA (2013) The effects of carbon dioxide and temperature on microRNA expression in *Arabidopsis* development. Nat Commun 4:2145
Medlyn BE, McMurtrie RE (2005) Effects of CO_2 on plants at different timescales. In: Ehleringer JR, Cerling TE, Dearing MD (eds) A history of atmospheric CO_2 and its effects on plants, animals, and ecosystems. Springer, New York, NY, pp 441–467
Michael TP, McClung CR (2003) Enhancer trapping reveals widespread circadian clock transcriptional control in Arabidopsis. Plant Physiol 132(2):629–639
Moore BD, Cheng SH, Rice J, Seemann JR (1998) Sucrose cycling, Rubisco expression, and prediction of photosynthetic acclimation to elevated atmospheric CO_2. Plant Cell Environ 21(8):905–915
Moore BD, Cheng SH, Sims D, Seemann JR (1999) The biochemical and molecular basis for photosynthetic acclimation to elevated atmospheric CO_2. Plant Cell Environ 22(6):567–582
Moore B, Zhou L, Rolland F, Hall Q, Cheng W-D, Liu Y-X et al (2003) Role of the Arabidopsis glucose sensor HXK1 in nutrient, light, and hormone signaling. Science 300:332–336
Nicotra AB, Atkin OK, Bonser SP, Davidson AM, Finnegan EJ, Mathesius U, Poot P, Purugganan MD, Richards CL, Valladares F, van Kleunen M (2010) Plant phenotypic plasticity in a changing climate. Trends Plant Sci 15(12):684–692
Norby RJ, Wullschleger SD, Gunderson CA, Johnson DW, Ceulemans R (1999) Tree responses to rising CO_2: implications for the future forest. Plant Cell Environ 22(6):683–714
Norby RJ, DeLucia EH, Gielen B, Calfapietra C, Giardina CP, King JS et al (2005) Forest response to elevated CO_2 is conserved across a broad range of productivity. Proc Natl Acad Sci U S A 102(50):18052–18056
Pauls SU, Nowak C, Bálint M, Pfenninger M (2013) The impact of global climate change on genetic diversity within populations and species. Mol Ecol 22(4):925–946
Pigliucci M (2001) Phenotypic plasticity: beyond nature and nurture. Johns Hopkins University Press, Baltimore, MD, p 333
Poorter H, Pèrez-Soba M (2002) Plant growth at elevated CO_2. The Earth system: biological and ecological dimensions of global environmental change. In: Mooney HA, Canadell JG (eds) Encyclopedia of global environmental change. John Wiley & Sons, Chichester, pp 489–496
Pritchard SF, Rogers HH, Prior SA, Peterson CM (1999) Elevated CO_2 and plant structure: a review. Glob Change Biol 5(7):807–837
Rae AM, Ferris R, Tallis MJ, Taylor G (2006) Elucidating genomic regions determining enhanced leaf growth and delayed senescence in elevated CO_2. Plant Cell Environ 29(9):1730–1741
Rae AM, Tricker PJ, Bunn SM, Taylor G (2007) Adaptation of tree growth to elevated CO_2: quantitative trait *loci* for biomass in *Populus*. New Phytol 175(1):59–69
Richardson BA, Shaw NL, Pendleton RL (2012) Plant vulnerabilities and genetic adaptation. In: Finch DM (ed) Climate change in grasslands, shrublands, and deserts of the interior American West: a review and needs assessment. U.S. Department of Agriculture, Forest Service, Rocky Mountain Research Station, Fort Collins, CO, pp 48–59
Seiler TJ, Rasse DP, Li JH, Dijkstra P, Anderson HP, Johnson DP et al (2009) Disturbance, rainfall and contrasting species responses mediated above-ground biomass response to 11 years of CO_2 enrichment in a Florida scrub-oak ecosystem. Glob Change Biol 15(2):356–367
Solomon S, Qin D, Manning M, Chen Z, Marquis M, Averyt KB et al (eds) (2007) Contribution of working group I to the fourth assessment report of the Intergovernmental Panel on Climate Change. Cambridge University Press, Cambridge

Springer CJ, Ward JK (2007) Flowering time and elevated atmospheric CO_2. New Phytol 176(2):243–255
Stitt M (1986) Limitation of photosynthesis by carbon metabolism. Physiol Plant 81(4):1115–1122
Stitt M (1991) Rising CO_2 levels and their potential significance for carbon flow in photosynthetic cells. Plant Cell Environ 14(8):741–762
Tallis MJ, Lin Y, Rogers A, Zhang J, Street NR, Miglietta F, Karnosky DF, De Angelis P, Calfapietra C, Taylor G (2010) The transcriptome of *Populus* in elevated CO_2 reveals increased anthocyanin biosynthesis during delayed autumnal senescence. New Phytol 186(2):415–428
Taylor G, Street NR, Tricker PJ, Sjödin A, Graham L, Skogström O, Calfapietra C, Scarascia-Mugnozza G, Jansson S (2005) The transcriptome of *Populus* in elevated CO_2. New Phytol 167(1):143–154
Thudi M, Li Y, Jackson SA, May GD, Varshney RK (2012) Current state-of-art of sequencing technologies for plant genomics research. Brief Funct Genomics 11(1):3–11
Vu JCV, Allen LH, Gesch RW (2006) Up-regulation of photosynthesis and sucrose metabolism enzymes in young expanding leaves of sugarcane under elevated growth CO_2. Plant Sci 171(1):123–131
Wall GW, Brooks TJ, Adam NR, Cousins AB, Kimball BA, Pinter PJ et al (2011) Elevated atmospheric CO_2 improved *Sorghum* plant water status by ameliorating the adverse effects of drought. New Phytol 152(2):231–248
Wand SJE, Midgley GF, Jones MH, Curtis PS (1999) Responses of wild C_4 and C_3 grass (Poaceae) species to elevated atmospheric CO_2 concentration: a meta-analytic test of current theories and perceptions. Glob Change Biol 5(6):723–741
Wang D, Heckathorn SA, Wand X, Philpott SM (2012) A meta-analysis of plant physiological and growth responses to temperature and elevated CO_2. Oecologia 169(1):1–13
Ward JK, Kelly JK (2004) Scaling up evolutionary responses to elevated CO_2: lessons from *Arabidopsis*. Ecol Lett 7(5):427–440
Ward JK, Roy DS, Chatterjee I, Bone CR, Springer CJ, Kelly JK (2012) Identification of a major QTL that alters flowering time at elevated [CO_2] in *Arabidopsis thaliana*. PLoS One 7(11):1–9
Watling JR, Press MC (1997) How is the relationship between C_4 cereal *Sorghum bicolor* and the C_3 root hemi-parasites *Striga hermonthica* and *Striga asiatica* affected by elevated CO_2? Plant Cell Environ 20(10):1292–1300
Watling JR, Press MC, Quick WP (2000) Elevated CO_2 induces biochemical and ultrastructural changes in leaves of the C_4 cereal sorghum. Plant Physiol 123(3):1143–1152
Wei H, Gou J, Yordanov Y, Zhang H, Thakur R, Jones W, Burton A (2013) Global transcriptomic profiling of aspen trees under elevated [CO_2] to identify potential molecular mechanisms responsible for enhanced radial growth. J Plant Res 126(2):305–320
Wong SC, Cowan IR, Farquhar GD (1979) Stomatal conductance correlates with photosynthetic capacity. Nature 282:424–426
Ziska LH, Bunce JA (1997) Influence of increasing carbon dioxide concentration on the photosynthetic and growth stimulation of selected C_4 crops and weeds. Photosyn Res 54:199–208
Ziska LH, Sicher RC, Bunce JA (1999) The impact of elevated carbon dioxide on the growth and gas exchange of three C_4 species differing in CO_2 leak rates. Physiol Plant 105(1):74–80

Genomics of Drought

Tiago F. Lourenço, Pedro M. Barros, Nelson J.M. Saibo, Isabel A. Abreu, Ana Paula Santos, Carla António, João S. Pereira, and M. Margarida Oliveira

Drought: Time to Adapt?

Plant Adaptation to Terrestrial Conditions

About 400 million years ago (by the end of the Silurian or early Devonian), some plants began colonizing emersed, well drained habitats. Since then, plants relentlessly occupied the whole planet. The major challenge was: display of leaves to do photosynthesis without losing too much water to a dry atmosphere. In this process, plants had to "learn" how to deal with variations in water availability (in time and in space). This included the development of a water-proof outer layer of hydrophobic polymers (suberins and cutins), together with specialized systems for gas exchange across the epidermis—the stomata. On the other hand, to supply leaves with water, required a vascular system capable of long-distance water transport and another very relevant cell wall compound—the impermeable lignin. Other changes happened to fix the plants to soil and to guarantee that roots can access the photoassimilates, as well as to provide a good distribution of water and nutrients through all under-ground and above-ground tissues. In this context, the vascular system was a major "invention" in the evolution process and its resistance and efficiency plays a major role in the way a plant deals with water variations.

T.F. Lourenço, Ph.D. (✉) • P.M. Barros, Ph.D. (✉) • N.J.M. Saibo, Ph.D. • I.A. Abreu, Ph.D.
A.P. Santos, Ph.D. • C. António, Ph.D. • M.M. Oliveira, Ph.D. (✉)
Instituto de Tecnologia Química e Biológica António Xavier, Universidade Nova de Lisboa, Av. Da República, Oeiras 2780-157, Portugal
e-mail: mmolive@itqb.unl.pt

J.S. Pereira, Ph.D.
Instituto Superior de Agronomia, Universidade de Lisboa, Lisbon, Portugal

D. Edwards, J. Batley (eds.), *Plant Genomics and Climate Change*,
DOI 10.1007/978-1-4939-3536-9_5

Mosses are probably among the first plants that were able to colonize the land. Although they appear as very simple organisms, they are also among the most drought tolerant terrestrial plants, efficiently shutting down their metabolism in a constitutive manner when drought arises, and fully recovering after some moisture. Bryophytes will not be covered in this revision, but the readers can find numerous references in the literature (see for instance: Oliver 2008; Koster et al. 2010; Stark et al. 2013).

Water Deficit and Plant Stress

It is considered that a plant is under stress when it faces sub-optimal conditions that may lead to the disruption of homeostasis in any metabolic pathway. It is obvious that water loss by a plant will translate into stress, as some processes or reactions will be affected by decreasing water potentials. Stress will be alleviated as uptake compensates for water losses.

During a plant drying event, with declining water availability, the first main biophysical effect of water deficit is the reduction of cell turgor pressure, and thus of cell wall extensibility, cell expansion and growth. Since soil drying leads to a decline in nutrient absorption (e.g.: nitrogen and calcium), some drought responses also allow plant adjustment to low mineral resources (McDonald and Davies 1996). McDowell et al. (2008) developed a hydraulically-based theory that helped analysis of hypotheses regarding plant survival and mortality under water stress. As reported by them, isohydric regulation of water status results from avoidance of drought-induced hydraulic failure, which is achieved by stomatal closure. Closed stomata lead to carbon starvation and to a cascade of effects such as increased susceptibility to biotic stress. Anisohydric plants, although showing some degree of drought-tolerance, are predisposed to hydraulic failure because they operate with narrower hydraulic safety margins during drought (McDowell et al. 2008).

Plants adapted to seasonally dry environments, such as desert plants, are able to avoid dehydration by restricting their life-cycle to ensure the completion of the full life cycle during the short period of water availability (Mooney et al. 1987; Maroco et al. 2000). Mediterranean herbaceous annuals behave similarly, spending the summer as seed, but perennial plants have to withstand periods of water shortage and acquired increased protection against water loss by depositing leaf wax, having sunken stomata (eg. *Pinus* or *Nerium* species) or small leaves, or going dormant and dropping leaves when approaching the dry season. These features also exist in succulent perennials such as cacti, some Euphorbiaceae or Asparagaceae. Succulence is given by a water storage tissue, composed of thin-walled cells, well protected by a thick epidermis and a water-impermeable waxy cuticle. The associated succulent metabolism also provides specific protection against water loss, since water deficit activates a metabolic change from the C3 to the Crassulaceaen Acid metabolism (CAM). This makes stomata change their regulation and start closing during day and opening at night (when temperatures are lower) to fix CO_2

in organic acids and later release it to feed the photosynthetic metabolism. To reduce leaf surface and transpiration area, some CAM plants, like some Cactaceae sp., also strongly reduce leaf size (to spines) or even don't develop them at all. A good recent revision of the morpho-anatomical adaptations to drought was made by De Micco and Aronne (2012).

Although stomata are the major regulators of plant gas exchange, plants may also succeed in reducing water loss by increasing reflectance through a dense trichome layer (Larcher 2000), or reducing the exposed leaf area and even reducing the canopy (Ehleringer and Cooper 1992) (see section "Drought Escape"). Leaf abscission may result from increased synthesis and perception of ethylene. Water stress reduces stomatal conductance in the older leaves, thus limiting their photosynthetic rate and fine-tuning the demand for resources. However, young leaves maintained during severe drought often show high rates of photosynthesis (Ludlow and Ng 1974) and Rubisco content, and often reach maturity and become carbon sources at a smaller size than leaves developing under well-watered conditions (Schurr et al. 2000). The slower formation of photosynthetic leaf area in turn reduces the flow of assimilates to the meristematic and growing tissues of the plant, fine-tuning the demand for resources.

Translocation, instead, is quite insensitive to drought, which allows seed germination in rather adverse environments. The way in which drought stress affects photosynthesis occurs at different levels: CO_2 diffusion, Photosystem II (PSII) efficiency, electron transport, ROS formation, RuBP content, Rubisco activity, and photorespiration (Saibo et al. 2009). Under drought conditions, plants try to maintain the photosynthetic efficiency as high as possible, but avoiding the overexcitation of the photosynthetic apparatus and consequent photooxidative damage (Chaves et al. 2011). An important aspect of the drought stress plant response is the recovery capacity after stress. The way photosynthesis recovers after re-watering largely determines plant tolerance to drought (Flexas et al. 2006; Chaves et al. 2009).

Photosynthesis is more tolerant to mild water deficits than cell expansion and leaf growth. Shoot growth inhibition under drought reduces the plant metabolic demands and mobilizes metabolites for the synthesis of protective compounds required for osmotic adjustment. The synthesis and accumulation of compatible solutes (like amino acids, sugar-alcohols or glycine-betaine) help maintain the water equilibrium in the cell. Also, the production of antioxidants and other stress-relieving agents may help plants to adapt to the stress by overcoming its harmful effects. The production of reactive oxygen species (ROS) is one of the problems arising from the disruption of homeostasis. Since ROS are produced as a consequence of several different stresses, this also justifies that many stress responses follow common pathways (Sagi and Fluhr 2006). ROS are generated when (after stomata closure) low CO_2 concentration within the leaf and impaired ATP synthesis lead to a reduced Calvin cycle activity, and thus to a reduction in the final electron acceptor $NADP^+$. Increased ROS induce oxidative stress and damage (in proteins, DNA, lipids), inactivating the photochemical reaction center of PSII and leading to photoinhibition. Eventually, the environmental stress inactivation of PSII could be the result of the inhibition of its photodamage repair, rather than a direct attack (Murata et al. 2007 and references therein).

The reduction in turgor pressure often affects leaves more than roots (although roots are particularly affected at the root tip level in dry soils) (Munns and Sharp 1993). This means that for a given water stress, roots may grow more than leaves, thus leading to a greater root/shoot ratio in stressed plants. Some species may develop extensive root systems and long roots able to reach the water stored below ground. For instance, the small alfalfa (*Medicago* sp.), although it typically has roots no longer than 3 m, was also found with roots of 40 m, and another legume species, the mesquite tree (*Prosopis* sp.) has been reported with 53 m long roots (Jensen and Salisbury 1972). This intensive root growth may be associated with reduced production of lateral roots. However in *A. thaliana*, the inhibition of lateral root development by drought is an adaption mechanism that results mainly from inhibition of elongation, since the number of lateral primordia per root is similar in drought and control conditions (Xiong et al. 2006).

All these strategies minimize the impact of water loss by the plant. However, plants such as the true xerophytes (such as cacti and some bromeliads) are able to loose water at levels that would compromise survival in other species. These are the drought-enduring non-succulent plants. Using a number of strategies, that include the accumulation of compatible solutes and increased protection against reactive oxygen species, these species are able to recover very fast when water is again available, resuming growth and all normal functioning shortly after rehydration.

Long-lived leaves may also develop anatomical and morphological features as protection mechanisms. Sclerophylly is a dominant leaf-form in woody plants under Mediterranean climates, for example, and has been described as an adaptation to seasonal water deficits (e.g. Oertli et al. 1990), or a consequence of low-nutrient soils (e.g. Beadle 1966), among other roles. Sclerophyllous leaves are usually smaller and thicker, with a low specific leaf area and low nitrogen concentration, providing protection to extreme climatic conditions. Small leaves are well adapted to the high light and high temperatures that prevail in arid regions, because their size permits increased heat dissipation (sensible heat) and an effective control of water loss by stomatal closure (Jarvis and Mcnaughton 1986). Many of the evergreen shrubs and trees in arid or semi-arid regions combine high solute concentration in living cells with sclerophylly, low photosynthetic capacity and stomatal conductance (Faria et al. 1998).

How to Ensure Xylem Function Under Drought

One important consequence of drought stress is xylem cavitation, which imposes an important limitation in water transport and, consequently, stomatal conductance and photosynthetic activity. Water ascends in plants by capillary action of cell walls, driven by the surface tension generated at evaporating surfaces of leaves, with xylem pressure more negative than that of vapour pressure water (according to the cohesion-tension theory; Dixon and Joly 1895). Under negative pressure, liquid xylem sap becomes metastable, a state which is maintained by the cohesion of water to water and adhesion of water to walls of xylem conduits (Tyree 1997). These conditions, however, make water transport inherently vulnerable to cavitation (i.e. the rapid phase change of liquid water to vapour) during normal diurnal gas exchange, causing embolisms through blocking of water transport in the affected xylem vessels (Tyree 1997; Steudle 2001). When plants face a drought condition, more negative pressure develops and air may be aspirated into functional conduits through porous sections in their walls. Once inside the conduit, these air bubbles may seed the phase change to vapour, causing the negative sap pressure to rise abruptly to near atmospheric. The gas bubble eventually expands and fills the conduit, producing an embolism as water is drained by the surrounding transpiration stream (Lens et al. 2013). Air seeding may even propagate to surrounding conduits through interconduit pits, when there is enough pressure difference between them (Brodersen et al. 2013).

Most plants are able to restore vessel functionality by refilling embolized vessels with water, making use of the high root pressure reached during the night or when transpiration decreases (Kaufmann et al. 2009; Cao et al. 2012; Brodersen and McElrone 2013). Additionally, in woody angiosperms embolism repair may also occur through an energy gradient established to allow water to flow from surrounding living parenchyma cells into the embolized vessel (Holbrook and Zwieniecki 1999; Sperry 2003; Brodersen et al. 2013). Mechanosensing of the energy released during cavitation (Salleo et al. 2008) or changes in the osmotic properties of the embolized vessel (Secchi et al. 2011; Secchi and Zwieniecki 2011) may act as trigger signals for this response. Several studies have reported a decrease in starch content in xylem parenchyma upon embolism formation supporting the hypothesis that sugars derived from starch hydrolysis might be transported out of parenchyma into embolized vessels, generating an osmotic driving force for refilling (Nardini et al. 2011; Brodersen and McElrone 2013). In poplar, embolism formation caused the upregulation of genes involved in carbohydrate metabolic pathways as well as genes encoding aquaporins and ion transporters (Secchi et al. 2011). Up-regulation of carbohydrate metabolism (mostly disaccharide metabolism) in embolized stems could promote the release of sucrose that would be used as the source of energy needed to support refilling through increased respiration, or as an osmoticum (Secchi et al. 2011; Brodersen and McElrone 2013). Build up of osmoticum in the vessel lumen would trigger water efflux, facilitated by the up-regulation of aquaporins (Brodersen and McElrone 2013). Furthermore, the decrease in starch content may turn parenchyma cells into strong sinks to phloem-induced unloading of sugars and water in the direction of embolized vessels, through ray parenchyma (Nardini et al. 2011).

Embolism resistance may vary widely among species, depending on structural aspects such as xylem anatomy or specific pit morphology, which may avoid or prevent the spread of drought-induced embolism (Lens et al. 2013). In addition, in the case of water deficit, embolism repair may also depend on the plant's ability to maintain sufficient hydration levels to allow refilling the embolized vessels. Repeated cycles of embolism formation and repair may also deplete starch reserves and disable the refilling mechanism through parenchyma cells after consecutive stress events. Embolism is particularly important in woody plants, being considered one of the major physiological factors leading to drought-induced mortality and forest decline (Choat et al. 2012). However, the precise physiological mechanisms underlying plant mortality are still poorly understood, although hypothesis related to carbon starvation, due to prolonged negative carbohydrate balance, and hydraulic failure leading to tissue desiccation are currently under debate (reviewed by McDowell 2011).

Drought Damage in Crop Production

By definition a drought will negatively affect crop production. In some regions, climatic changes increased the frequency and severity of droughts, thus altering the existing conditions, affecting rain fed systems such as forests and limiting the irrigation potential for agriculture. Moreover, most crops have been bred for higher productivity instead of stress tolerance, although water availability is becoming a major problem not only in countries where food security is not guaranteed, but also in other regions using rainfall to sustain agriculture. It is for instance the case of the extensive wheat production in Australia, which is highly vulnerable to drought (Eisenstein 2013).

The impact of drought or limited water is expressed not only at the level of plant survival, but also of plant performance, since good growth and seed set, are imperative in agriculture. This makes water use efficiency particularly important, and the strategies that plants most efficiently use to deal with drought and to maximize water, one of the major research targets. Below we review some strategies that better enable plants to deal with water limiting conditions, detailing the molecular aspects orchestrating their responses.

The Main Strategies of Plant Response to Drought

Plant responses to drought involve both phenological and physiological adjustments, which are defined by the water status of both plant and environment. Plants can either be subjected to slowly developing water shortage (occurring within days, weeks or months) or face short-term water deficit (occurring within hours to several days). In the first case, plants adapt to these conditions by either shortening their life cycle or by optimizing their resource gain in the long term through acclimation responses. In case of sudden dehydration, plants tend to minimise water loss or exhibit metabolic protection against the damaging effects of dehydration and developing oxidative stress (Chaves et al. 2003). The physiological adaptations to drought are to a great extent determined by the physical consequences of stomata closure: the altered trade-off between the CO_2 uptake supporting growth and water loss due to transpiration. The efficiency of photosynthetic carbon gain relative to the rate of water loss (water use efficiency, WUE) can be used as an indicator of the ability to which plants resolve this trade-off in response to environmental changes. The importance of stomatal regulation and photosynthetic efficiency also deserved a special focus within dedicated sections (see section "Regulation of Stomatal Development and Aperture" and "Regulation of Photosynthesis"). Plants that adapt to drought conditions are able to avoid tissue dehydration and maintain high tissue water potential through an increase in WUE (avoidance), or able to tolerate lower tissue water potential (tolerance).

The physiological strategies of drought stress response are classically grouped in drought escape and drought resistance, the latter being further divided in drought avoidance and drought tolerance (Levitt 1980). These strategies are not mutually exclusive (Ludlow 1989) and likely play differing roles across species and stress conditions, depending on duration, intensity or timing (Juenger 2013). We have further used the terminology desiccation tolerance to encompass those few higher plants able to survive complete desiccation and rehydration. In Fig. 1 we present a comprehensive scheme of the different strategies that higher plants use to deal with water deficit. The morphophysiological and molecular mechanisms that define these different strategies do not have well-defined frontiers and different plants combine various mechanisms, especially in what concerns reduction of water loss, cellular protection (against ROS and excessive photosynthesis) and accumulation of compatible solutes.

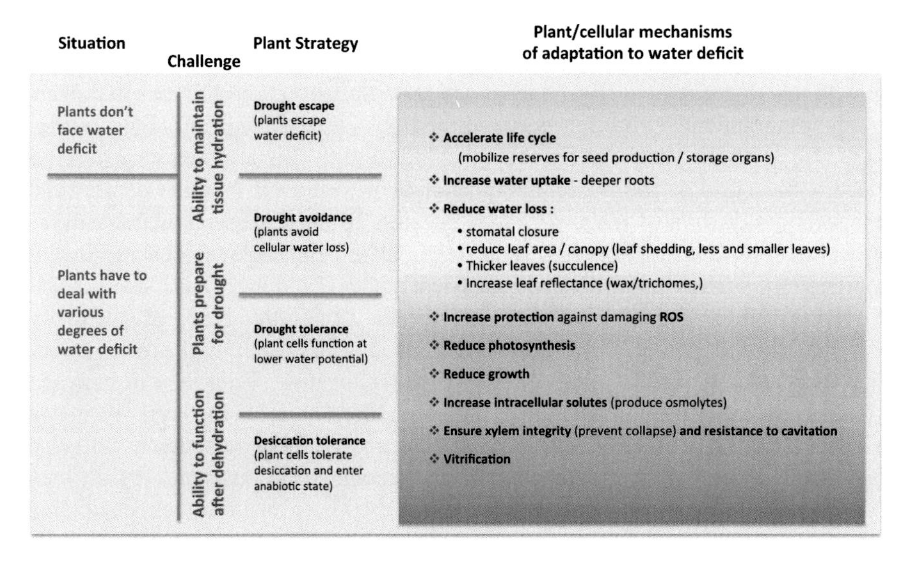

Fig. 1 Plants have different ways to deal with water limitations. They may avoid tissue dehydration using drought escape or drought avoidance strategies, or they may develop a number of mechanisms that confer them varying degrees of protection against water loss, as is the case of drought tolerance or desiccation tolerance. The different adaptation mechanisms may occur in different plants or in the same plant. See text for further explanations

Drought Escape

Escaping drought occurs when plants complete their life cycle before the most intense period of drought, through an increase in metabolic activity, accelerated growth and early flowering. This shift in development is supported by an increase in photosynthetic and gas exchange rate, which allows them to rapidly gain carbon as they grow and develop at a cost of a decrease in WUE (Franks 2011). Besides early flowering, an increased speed of fruit maturation is also crucial to guarantee reproduction success (Wu et al. 2010). This may be achieved by a better partitioning of pre-stored assimilates to developing seeds, which was already documented for some cereals (Yang et al. 2000, 2001) and legumes (Rodrigues et al. 1995) under water deficit. In this way, plants skip the negative effects of drought, using the maximum available resources while soil moisture lasts, in a 'live fast, die young' manner (Chaves et al. 2003; Wu et al. 2010).

Drought Avoidance

Drought avoidance is accomplished through the decrease in water loss, generally by the regulation of stomatal conductance (Chaves et al. 2003) (see section "Drought-Affected Developmental Traits"), and by an increase in water uptake through an investment in root system (Poorter et al. 2012) (see section "Regulation of Root Traits"). A tight control of stomatal conductance will directly hinder photosynthetic assimilation (given the decrease in CO_2 uptake), limiting metabolic activity and growth rate (Schulze et al. 1987; Geber and Dawson 1997). The decrease in the rate of photosynthetic electron transport and the light-saturated rate of photosynthesis may also increase plants susceptibility to high light. This is potentially harmful since it may lead to the photoinhibition of photosystem II reaction centres and enhanced production of reactive oxygen species. To avoid these consequences, as water stress develops, plants may reduce light absorbance through changes in leaf morphology, such as leaf rolling and steep leaf angles (Ehleringer and Cooper 2006; Kadioglu et al. 2012), reduction of the specific leaf area (ratio of the leaf area to dry weight) (Liu and Stützel 2004), diminishing canopy leaf area through growth reduction and shedding of older leaves (allowing the reallocation of nutrients to stem and younger leaves) or the construction of a highly reflective trychome layer (Bosu and Wagner 2007; Galmes 2007). Besides light absorbance, these adaptations will also play a role in decreasing leaf temperature and reducing transpiration.

Some plant species may potentially adapt to escape or avoid drought depending on site-specific environmental conditions. Many studies comparing natural plant populations adapted to contrasting environments suggest the occurrence of trade-offs between both strategies (e.g. McKay et al. 2003; Wu et al. 2010; Franks 2011). Thus, while a more conservative water use could allow the plant to extend its life cycle over a longer period, under a water-limiting environment it may reduce the rate of growth and development and constrain or prevent drought escape. In this scenario, natural selection would tend to favour one strategy over the other, depending on the interaction between environmental conditions and the life history of the plant (Heschel and Riginos 2005). In seasonally dry and variable environments, such as the Mediterranean climate, the strategy of a decreased WUE during periods of abundance, in order to escape drought by completing the life cycle faster, showed to be a successful strategy in specific cases (McKay et al. 2003; Heschel and Riginos 2005; Sherrard and Maherali 2006; Franks 2011). However, in environments that are consistently water-limited, drought avoidance can be an adaptive trait (Ludlow 1989).

Drought Tolerance

Plants can also endure drought conditions by tolerating low tissue water potential through osmotic adjustment or specific aspects of cell structure such as smaller size or rigid cell walls (Chaves et al. 2003). Drought tolerant plants can survive moderate dehydration, down to a moisture content below which there is no bulk cytoplasmic water present (~23 % water on a fresh weight basis, or ~0.3 g H_2O g^{-1} dry mass) (Hoekstra et al. 2001). Water is the driving force for the assembly of biological membranes and, in part, for the conformation of many proteins. Water loss will lead to a reduction of intracellular volume and increase of viscosity, which will crowd the cytoplasm and increase the chance of protein denaturation and membrane fusion. By osmotic adjustment, plants are able to decrease osmotic potential by the net increase of intracellular solutes, allowing the maintenance of turgor and sustaining metabolic activity and other processes (such as cell enlargement and stomatal control) after the onset of drought (Zhang et al. 1999). A broad range of organic compounds has been identified that can prevent adverse molecular interactions during drought, which include proline, glutamate, glycine-betaine, mannitol, sorbitol, fructans, polyols, trehalose, sucrose as well as raffinose family oligosaccharides (RFOs) (Hoekstra et al. 2001). Section "Accumulation of Osmolytes/Compatible Solutes" in this chapter reviews some of these compounds, the molecular control of their production and putative modes of action.

Desiccation Tolerance

If moisture levels inside the cell drop to 10 % absolute water content or less, the hydration shell of molecules is gradually lost (Hoekstra et al. 2001; Le and McQueen-Mason 2007). This extreme loss of water or desiccation is tolerated only by seeds, some pollen grains and by a small group of the so-called resurrection plants, which are able to withstand months to years without water (Gechev et al. 2012). Most resurrection plants counteract developing water deficiency using response mechanisms similar to other drought-tolerant plants. Yet, many of these protective mechanisms were shown to be more substantially activated in resurrection plants upon sensing water deficiency (Djilianov et al. 2011; Gechev et al. 2012). Under drought, resurrection plants may further lose most of the free water in their vegetative tissues, falling into anabiosis but, upon rewatering, they quickly regain normal activity. Up to now about 1300 resurrection plant species have been reported, 1000 among pteridophytes (e.g. ferns and mosses) and 300 angiosperms (Porembski 2011).

The ability of some resurrection plants to tolerate the desiccation conditions is in great part associated to the large accumulation of sugars (such as sucrose and trehalose), which can structurally and functionally protect proteins and membranes by replacing water. Sugars are also important in vitrification of the cell cytoplasm during desiccation, i.e. the formation of biological glasses within the dried cell which may decrease chemical reaction rates and molecular diffusion, and also limit oxidative damage (Hoekstra 2005; Le and McQueen-Mason 2007). Oxidative stress damage is further reduced by the induction of anti-oxidant enzymes and the inhibition of photosynthesis. In fact, the antioxidants and other protective molecules accumulated can sometimes reach more than 70 % of plants dry weight, as reported for *Myrothamnus flabellifolius*, that exists in a dehydrated quiescent state for approximately half of the year (Moore et al. 2005). Another adaptation to desiccation in resurrection plants involves the increase in cell wall flexibility. This is mediated by the induction of expansins, non-enzymatic proteins which loosen plant cell walls, minimizing mechanical damage due to vacuolar shrinking (Le and McQueen-Mason 2007). Yet desiccation tolerance in resurrection plants is not solely characterized by their response during actual desiccation. This is because some of these plants appear to be already primed for desiccation, accumulating higher levels of transcripts or proteins related to cellular protection (e.g. antioxidants, hydrophilins, cell remodelling proteins) even in favourable (non stress) growth conditions (Oliver et al. 2011; Gechev et al. 2012). This aspect may come with the expense of reduced growth rates, since part of the energy resources are directed towards synthesis of stress-protective compounds (Gechev et al. 2012).

Dormancy: A Helpful Strategy

Although it doesn't seem to be triggered by drought, the establishment of a dormancy state may be an important drought endurance strategy, involving the suppression of growth and other physiological activities during periods of extreme conditions. This is particularly important in a seasonal context, as reported for several Mediterranean geophytes (the name given to perennial plants propagated by buds existing on underground bulbs, tubers, or corms) (Alliaceae, Orchidaceae, Poaceae and Liliaceae) (Volaire and Norton 2006). Summer dormancy in these species can be characterized by the reduction or cessation of leaf production and expansion, senescence of mature foliage, dehydration of surviving organs, and in some species, the formation of resting organs (Volaire and Norton 2006). The mechanisms by which Mediterranean perennial grasses ultimately survive extreme drought conditions are to some extent similar to what is described for drought avoidance or tolerance in other plant species (see sections "Drought Avoidance" and "Drought Tolerance"). For example, different cultivars from tall fescue (*Lolium arundinaceum*) may develop only partial dormancy, showing reduced leaf senescence and greater water content in basal tissues, which is supported by high root length densities (Lelievre et al. 2011). However, drought tolerance and seasonal dormancy induction seem to be independent phenomena (Volaire and Norton 2006; Volaire et al. 2009), since dehydration tolerance may be induced when these plants are subjected to drought at any time of the year. In turn, dormancy is only exhibited in summer since it only develops according to specific environmental cues such as increasing daylength and temperature. The suspension of most metabolic processes during dormancy will also prevent regrowth in the event of occasional summer rains, as this might be detrimental to summer survival (Volaire and Norton 2006). This is one of the main differences between desiccation tolerance in resurrection plants and summer dormancy in perennial grasses.

A Molecular Picture of Early Events in Drought Response

To cope with water scarcity, some plants have evolved complex mechanisms for rapid adaptation to such adverse environments that involve extensive molecular reprogramming at the transcriptional, proteomic and metabolomic levels. Stress recognition activates signal transduction pathways that transfer information within the cell and throughout the whole plant. Unlike the submergence responses in rice, which are mainly controlled by the *SNORKEL* and *SUBMERGENCE-1* loci, drought response is not controlled by a single master regulator (Fukao and Xiong 2013). This is very clear from the different examples of variable plant responses to water deficit. In fact, although several quantitative trait loci (QTL) have been identified in several crops such as rice, maize, barley, wheat or sorgum, they can only explain a small part of the phenotypic variation (Mir et al. 2012).

Designing the Experiment: Stress and Recovery

To investigate the extensive molecular reprogramming occurring upon drought and correlate it with physiological parameters, several studies have been performed using different species (*Arabidopsis*, rice, barley, grapevine, *Pinus*, *Populus*, maize), different samples and tissues (whole plant, shoot, root, seeds, xylem sap), as well as different developmental stages, among other variants. The experiments were conducted using several different approaches, including microarrays and state-of-the-art RNAseq (Ozturk et al. 2002; Seki et al. 2002; Oono et al. 2003; Rabbani et al.

2003; Ueda et al. 2004; Bogeat-Triboulot et al. 2007; Cramer et al. 2007; Alvarez et al. 2008; Zeller et al. 2009; Harb et al. 2010; Liu and Jiang 2010; Deokar et al. 2011; Lorenz et al. 2011; Yan et al. 2012; Jin et al. 2013; Xu et al. 2013). Nevertheless, the comparison of all the data is very difficult, even within the same species, due to the different methodologies used, and experimental conditions applied, including variations in stress severity and duration, or statistical analysis. An additional relevant factor is that the conditions assayed in the lab, hardly reproduce what happens in the field, and good results in the lab, may not transpose as efficiently to the field conditions.

In spite of all the factors that complicate the comparison of the different experiments and data, the classes of up- or down-regulated genes are similar, probably indicating that the molecular mechanisms of response to water deficit were conserved along evolution (Liu and Jiang 2010; Lorenz et al. 2011). Among the more represented classes of genes are regulatory proteins involved in signal transduction, regulating the metabolism of abscisic acid (ABA) and phospholipids, or encoding transcription factors, kinases, phosphatases, and proteins such as the late embryogenesis abundant (LEAs), the small heat shock proteins (HSP) and chaperones, water channels, enzymes for the biosynthesis of osmolytes (compatible solutes) and for detoxification of reactive oxygen species (ROS) (Shinozaki and Yamaguchi-Shinozaki 2007; Hirayama and Shinozaki 2010). The plant capacity to recover from stress is particularly relevant (Lorenz et al. 2011; Almeida et al. 2013; Sapeta et al. 2013) although it is not commonly analysed in most molecular studies. A recent study focusing on *Jatropha curcas*, an emerging bio-diesel crop, has shown that, after a 28 day period of water withhold (15 % soil water capacity), this plant is able to fully restore the photosynthetic function within the first 48 h of re-hydration, showing normal values for parameters such as PSII efficiency (A_n, g_s,) (Sapeta et al. 2013). The molecular basis of this fast recovery was also targeted in a transcriptomic study by RNAseq (Sapeta et al. 2015). Using a similar approach, a transcriptomic study of *Pinus taeda* roots subjected to drought and subsequent recovery has shown a fast recovery of the gene expression pattern to pre-drought levels, within 48 h of re-hydration (Lorenz et al. 2011). However, in pine, the authors assessed water stress by measuring plant water potential, while for *Jatropha curcas*, the soil water content was used as stress reference.

Drought Perception and Signalling

Signals for water stress (chemical or hydraulic) are produced, sensed and transduced by the plant into cellular responses resulting in physiological changes that express the drought stress response. Like in other abiotic stresses, little is known about the early events in the perception of drought signals. AtHTK1, a transmembrane histidine kinase, was shown to be upregulated in response to water deficit and suggested to be one of the specific receptors perceiving drought, capable of activating a downstream signaling cascade (Urao et al. 1999).

It is expected that water deficit/changes in osmotic status result in changes in the physical state of membranes and thus in protein-lipid interactions. Thus, transmembrane proteins like ionic channels, kinases or water channels (aquaporins), are good candidates to be the first to perceive such signals. Their gene expression may be regulated in response to drought, but post-transcriptional and post-translational processes are emerging as of crucial importance in their regulation. As an example, aquaporins that play a central role in controlling water fluxes during drought (for a review see Maurel et al. 2008), were shown to be mostly down-regulated in drought-stressed plants, in species ranging from Arabidopsis to trees (Jang et al. 2004; Mahdieh et al. 2008; Almeida-Rodriguez et al. 2010; Yan et al. 2012). Although some specific aquaporins can be up-regulated in certain drought stress regimes (Yan et al. 2012), in *Mesembryanthemum crystallinum* (the common ice plant) they appear to be mostly regulated at post-transcriptional and post-translational levels (Kirch et al. 2000; Tyerman et al. 2002; Chaves et al. 2003).

The common drought signaling pathways have been distinguished as abscisic acid (ABA)-dependent and -independent (Shinozaki and Yamaguchi-Shinozaki 1997, 2007). The two paths were initially identified in Arabidopsis (the model plant for dicots) and later confirmed in rice (model plant for monocots) (Nakashima et al. 2009), showing that they are transversal among plants.

The Alternative Oxidase (AOX)

The alternative oxidase (AOX) that plant mitochondria have (as opposite to animals), is a relevant non-energy conserving branch of the electron transport, avoiding the generation of ROS chain at mitochondrial level (for a review on AOX in stress response see Vanlerberghe 2013). Wand and Vanlerberghe (2013) verified that in tobacco, the lack of mitochondrial AOX compromised recovery from severe drought stress, while mild to moderate drought slightly increased AOX accumulation, in parallel with progressive increase of several ROS scavenging components.

Interestingly, in durum wheat, several systems can dampen mitochondrial ROS production: the ATP-sensitive plant mitochondrial potassium channel ($\text{PmitoK}_{\text{ATP}}$); the plant uncoupling protein (PUCP); and the alternative oxidase (AOX), although only AOX is not activated by ROS in either control or stress conditions (Pastore et al. 2007). Still in wheat, Bartoli et al. (2005) demonstrated that under drought, the respiratory AOX pathway is up-regulated, thus increasing protection of the photosynthetic electron transport chain from excessive light damage.

In rice seedlings, drought induces AOX1a and AOX1b transcription expression and Feng et al. (2009) found that previous AOX inhibition lowers relative water content in the seedlings, thus pointing for a role of the AOX pathway in drought tolerance.

Role of Secondary Messengers (Calcium and ROS)

Secondary messengers, such as Ca^{2+} and reactive oxygen species (ROS), play an important role in activating signaling events. Several Ca^{2+}-dependent (directly or via interaction with calmodulin or CBL proteins) protein kinases were shown to be important in drought response. Among these are rice OsCIPK12, a CBL-interacting protein kinase, and OsCDPK7, a Ca^{2+}-dependent protein kinase (Saijo et al. 2000; Xiang et al. 2007). A rice Mitogen-Activated Protein Kinase Kinase Kinase (MAPKKK), DSM1, was recently suggested to be involved in ROS signaling, while playing an essential role in drought stress (Ning et al. 2010). Mitogen-Activated Protein Kinase (MAPK) cascades play an important role in mediating stress responses (for recent review see Sinha et al. 2011). Several components of these cascades were shown to respond to water-stress, such as alfalfa MAPKK p44MKK4 whose expression and activity are increased under drought in an ABA-independent way (Jonak et al. 1996), but only a few were proven to elicit actual physiological responses. As examples, in rice, OsMAPK5 was found to positively regulate drought tolerance (Xiong and Yang 2003), and overexpression of DSM1 increased tolerance to dehydration (Ning et al. 2010). ROS play a pivotal role in regulating plant biotic and abiotic stress responses (for recent review see Baxter et al. 2013). In particular, hydrogen peroxide (H_2O_2) is involved in long-distance signaling induced by several abiotic stresses (heat, cold, high-intensity light, and salinity stresses) and this signaling is dependent on respiratory burst oxidase homolog D (RbohD) gene (Porembski 2011). RBOHD is a NADPH oxidase and these oxidases are known to play a key role in plant ROS production (for recent review see Suzuki et al. 2011). Recently, ROS were proposed to mediate ABA-signaling in Arabidopsis guard cells, regulating stomatal response to ABA (Munemasa et al. 2013). Several other protein kinases have also been shown to play critical roles in rice drought tolerance, such as calcineurin B-like protein interacting protein kinase (CIPK) (Xiang et al. 2007), calcium dependent protein kinase (CDPK or CPK) (Saijo et al. 2000), and receptor-like kinase (Ouyang et al. 2010).

ABA-Dependent Response

Abscisic acid accumulation occurs immediately after drought and it is perceived by ABA receptors of the PYR/PYL/RCAR family that inhibit protein phosphatase 2C (PP2C), in an ABA-dependent manner (Ma et al. 2009; Park et al. 2009). Since, PP2C are negative regulators of ABA-responsive SNF1-related protein kinases 2 (SnRK2s) in the presence of ABA, PP2C inhibition releases SnRK2 proteins for activation (Umezawa et al. 2009; Vlad et al. 2009). Activated SnRK2s phosphorylated substrate proteins involved in stress response and developmental programs (Fig. 2), PYR/PYL/RCAR, PP2Cs and SnRK2s, are thus the core constituents of ABA-dependent signaling and were shown to be sufficient to maintain

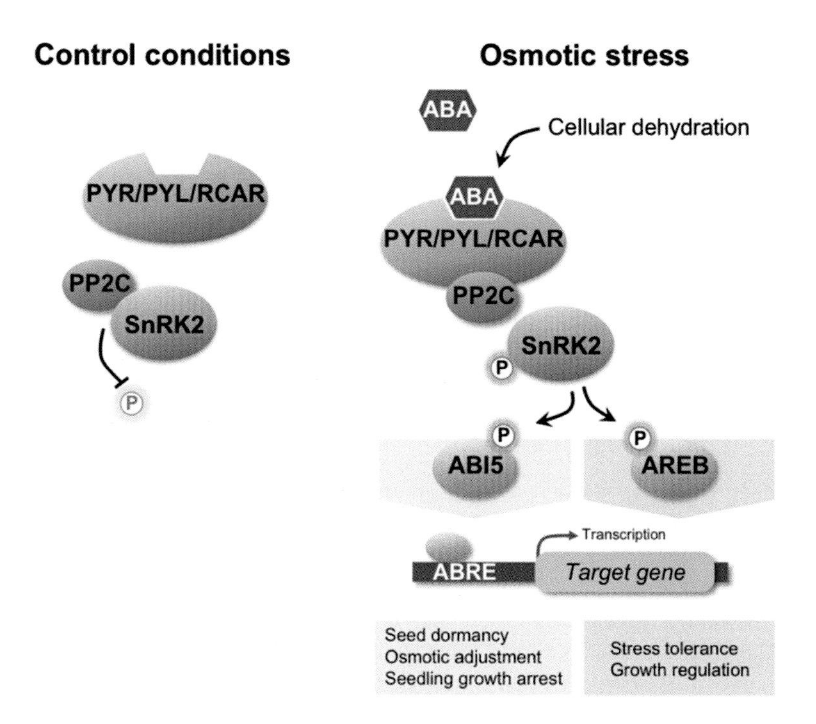

Fig. 2 Schematic representation of the ABA-dependent signaling pathway. Under control conditions (low ABA), the phosphatase PP2C interacts with the SnRK2 and represses its kinase activity. Under conditions of cellular dehydration associated either with seed maturation or drought stress (high ABA), the ABA-receptors PYR/PYL/RCAR recruit the phosphatase PP2C and release the SnRK2 to phosphorylate the downstream targets involved in ABA-response such as the bZIP transcription factors ABI5 or AREB/ABF. The different types of transcription factors are involved in ABA-dependent responses in seeds (ABI5) or vegetative tissue (AREB/ABF)

ABA-signaling in an *in vitro* reconstitution experiment (Fujii et al. 2009). Some ABA-dependent SnRK2s can also be directly activated by osmotic stress, such as SnRK2.6 (Boudsocq et al. 2004; Yoshida et al. 2006). Additionally, some SnRK2s are ABA-unresponsive, showing that the regulation of this kinase family is very complex. Their activity is regulated by phosphorylation, but although phosphorylation of specific residues is essential for SnRK2 *in vivo* activity, their constitutively-phosphorylated mimics (using acidic amino acid residues, such as Asp or Glu) does not result in constitutively-active kinases *in vivo*. This suggests that phosphorylation governs not only protein activity, but also other aspects, such as cellular localization or conformational changes (Fujii and Zhu 2012). The ABA-dependent pathway targets members of the ABA-responsive element (ABRE)-binding factors (ABFs) and these are well-characterized targets for SnRK2s. ABFs bind to the promoters of ABA-inducible genes to activate transcription. Other targets of SnRK2 kinases are ion-channels (see section "Regulation of Stomatal Development and Aperture"), NADPH oxidase in guard cells (thus regulating ROS signaling) enzymes involved in metabolic reactions, and dehydrins, accounting for their range of physiological effects (for a review on SnRK targets see Fujii and Zhu 2012).

Transcriptomic analyses have shown that thousands of genes are up- or down-regulated in response to drought and ABA and among these are genes associated with signal transduction. In these and other studies, several other protein kinases have also been shown to play critical roles in rice drought tolerance, such as calcineurin B-like protein interacting protein kinase (CIPK) (Xiang et al. 2007), calcium dependent protein kinase (CDPK or CPK) (Saijo et al. 2000), and receptor-like kinase (Ouyang et al. 2010).

ABA-Independent Response

ABA-independent responses can be mediated by the DRE/CRT *cis*-acting elements to which the transcription factors of the ERF/AP2 family bind, such as CBF/DREB1 and DREB2 (see details in section "Transcription Regulation of Drought Responses"). Among these, DREB2 genes respond to dehydration stress (Liu et al. 1998). Recently, it was suggested that lipid-mediated signaling is involved in *DREB2* regulation in Arabidopsis suspension cells and seedlings (Djafi et al. 2013). Phosphoinositide-dependent phospholipase C (PI-PLC) mediates the production of inositol triphosphate and diacylglycerol that can be further phosphorylated into phosphatidic acid. The basal activity of PI-PLC appears to result in the negative regulation of *DREB2A* expression (Perera et al. 2008; Djafi et al. 2013). Additionally, ABA-independent dehydration response can also be mediated by NAC transcription factors that bind the *cis*-elements in the promoter of the ERD gene. Little is known about the signaling that leads to NAC activation, but phosphorylation cascades must be involved, resulting in direct or indirect activation of these TFs. As an example, OsNAC4 translocation into the nucleus upon biotic stress is mediated by phosphorylation (Kaneda et al. 2009).

Molecular Mechanisms in Drought Stress Responses

As mentioned above, since not all genes have their expression altered by both ABA and drought, researchers have identified and distinguished the ABA-dependent and the ABA-independent signaling pathways, and invested in identifying the key genes involved (Shinozaki and Yamaguchi-Shinozaki 2007). In the past few years, attention was also given to other mechanisms controlling the responses to abiotic stress—the post-transcriptional regulation through miRNAs (Ding et al. 2013) and the regulation of the pool of key proteins by the ubiquitin/26S proteasome (Lyzenga and Stone 2012). In this chapter we review these aspects.

Hormonal Control of Drought Response: ABA and Hormone Interactions

ABA is a particularly important phytohormone in the response to water scarcity, inducing several physiological changes like reducing the transpiration rate by inducing stomatal closure (Himmelbach et al. 2003) (see section "Regulation of Stomatal Development and Aperture"), or controlling the plant daily water status oscillations through ABA-modulated gene expression related to the circadian clock (*TOC1*) (Legnaioli et al. 2009; Sanchez et al. 2011). The regulation of gene expression controlling guard cells movement is known to involve both ABA-dependent and ABA-independent signalling (Cominelli et al. 2005; Liang et al. 2005).

Among plant hormones, ABA shows the higher percentage of gene expression responsiveness (nearly 10 % of protein coding genes) (Fujita et al. 2011). ABA is also an important factor in the regulation of other processes such as seed development and maturation (Fujita et al. 2011). ABA is initially synthesized in plastids and the key enzymes of its synthesis and catabolism have been identified in *Arabidopsis* and other species (Kushiro et al. 2004; North et al. 2007; Wasilewska et al. 2008). ABA can circulate in the plant in an inactive form when conjugated in a glucose ester (ABAGE). To become active, b-glucosidases of apoplastic or endoplastic nature can break this glucose ester form to release active ABA.

It is commonly assumed that ABA is synthesized in roots upon osmotic stress and then translocated to leaves where it induces several molecular and physiological responses (e.g. stomata closure). However, there are reports that show that ABA is relocalized from leaves to roots upon drought stress (Christmann et al. 2005). These findings raise the question if ABA is in fact the drought signal translocated from roots to shoot. The hypothesis of another molecule or an hydraulic signal to be generated in roots and mobilized to the shoot to activate ABA production is debatable (Christmann et al. 2007; Wasilewska et al. 2008; Seo and Koshiba 2011).

Despite the importance of ABA, other phytohormones may also have a role in the responses to water deficit. An interesting feature controlling drought responses is the hormonal balance between ABA and jasmonic acid (JA). Several transcriptomic studies have revealed the modification of expression in genes related to JA synthesis and signaling under drought stress (Harb et al. 2010; Lorenz et al. 2011; Sapeta et al. 2015). It has long been known that jasmonates are implicated in drought related processes like stomatal closure (Creelman and Mullet 1995; Raghavendra and Reddy 1987). This role influencing stomatal closure was further observed in JA-insensitive mutants that could still respond to ABA-induced stomatal closure (Suhita et al. 2004) although to a lesser extent than the wild-type. An opposite, but similar, phenotype could be observed with the ABA mutant *ost1* which led to the conclusion of a possible crosstalk between these two plant hormones (Suhita et al. 2004). This crosstalk is putatively controlled through the action of a bHLH TF, MYC2, an important TF for the JA signaling that was found to also activate ABA-responsive genes (Hirayama and Shinozaki 2010; Lorenzo and Solano 2005).

Arabidopsis JA-insensitive mutants (*coi1* and *jin1*) have also been used in drought stress studies. These mutants were found to be impaired in growth arrest upon drought stress, leading to a biomass accumulation similar to that of well-watered plants (Harb et al. 2010). These observations led the authors to conclude that JA (or its signaling pathway) is needed to start the growth arrest program under drought conditions. It is probable that the combined action of ABA and JA (usually in negative correlation) is essential to control adaptation/acclimation to water limiting conditions and to resume growth afterwards.

Nevertheless, and despite the importance of ABA and JA balance to control drought responses especially at stomatal closure, other plant hormones may also influence drought responses (e.g. gibberellins, auxins in DRO1 function, section "Regulation of Root Traits"; brassinosteroids, section "Regulation of Stomatal Development and Aperture").

Transcription Regulation of Drought Responses

The ABA-Dependent Pathway

In the promoter regions of genes responding to ABA, a common *cis*-regulatory element essential to bind transcription factors (TFs) was found. This region (PyACGTGG/TC), named ABA-responsive element (ABRE), has to be present in multiple motifs or conjugated with a coupling element (CE). The TFs binding to it were designated as ABRE binding proteins (AREBs) or ABA binding factors (ABFs). These TFs, belonging to the bZIP family, were identified as modulated by ABA and responsible for the ABA-mediated modulation of gene expression. The members of the bZIP family involved in ABA responses belong to Group A. From the 13 members of this group, nine were shown to be involved in ABA-mediated responses. These TFs share the same gene organization with 3 N-terminal conserved regions and the bZIP-domain located in the C-terminal. This gene organization was shown to be important since members of group A lacking this gene organization are not involved in ABA-responses (Wasilewska et al. 2008; Fujita et al. 2011). Among the bZIP TFs involved in ABA responses, and especially in *Arabidopsis*, researchers were also able to distinguish between the ones involved in seed development and maturation, and the ones involved in osmotic stress responses. For instance, *ABI5* is involved in modulation of ABA-responses related to development (seed maturation) while the AREBs/ABFs (*AREB1*, *AREB2* and *ABF3*) are involved in osmotic stress responses. All these TFs are downstream and regulated by the PYR/PYL/RCAR-PP2C-SnRK2 pathway. The bZIP TFs that regulate ABA-responses to osmotic stress (AREB1, AREB2 and ABF3) require ABA for their full-activation and are able to form homo and hetero-dimers (Yoshida et al. 2010) and interact with SnRK2. Interestingly, these TFs have overlapping functions as their triple mutant is more sensitive to drought and resistant to ABA than the single or double mutants (Yoshida et al. 2010). Downstream of these TFs, and in response to drought stress, ABA signal causes a growth reduction leading to the repression of genes related to chloroplast and cell wall metabolism, ribosomal and plasma membrane. On the opposite side, genes related to regulatory functions like TFs, kinases and phosphatases, genes coding for enzymes involved in the synthesis of osmoprotectants, LEAs (Late embryogenesis abundant) proteins, carbohydrate metabolism, are up-regulated. Despite the importance of bZIP TFs in the modulation of drought responses through the ABA signal, other families of TFs have also been shown to mediate ABA-responses. Members of the AP2/ERF family, which includes members from the ABA-independent pathway, have been shown to mediate ABA-responses either as activators (DREB1A or DREB2C) or repressors (ABR1) (Pandey et al. 2005; Lee et al. 2010). Another class of TFs, the MYB TFs, were also shown to be involved in the activation of ABA-responsive genes (such as *RD22*) and affect the crosstalk between ABA and auxins (MYB2 and MYB96) (Fujita et al. 2011). Members of the NAM, ATAF, CUC (NAC) TFs were also shown to be regulated by ABA and drought and modulating downstream responses together with zinc-finger homeodomain TFs (ZF-HD). However, and similarly to AP2/ERF TFs, members of these TF families are also involved in the ABA-independent pathway responses to osmotic stress (Lata and Prasad 2011; Mizoi et al. 2012).

The ABA-Independent Pathway

Responses to drought/osmotic stress in an ABA-independent pathway arose with the identification, in *Arabidopsis*, of a *cis*-element in the promoter of the gene *RD29A*, which is induced by drought and ABA (Yamaguchi-Shinozaki and Shinozaki 1994). It was found that even in mutants for ABA synthesis or signaling, the *RD29A* gene could still respond to drought, leading to the possibility of an ABA-independent path. This *cis*-element was named dehydration responsive element (DRE, A/GCCGAC—core motif) and it is similar to the C-repeat responsive element (CRT) that is found in the promoter of several cold-responsive genes (Baker et al. 1994). The TFs binding to these regions belong to the AP2/ERF family and were designated as DRE binding proteins (DREB) or CRT-binding factor (CBF). Therefore, it is not surprising that different TFs of the DREB/CBF subfamily may modulate responses to either drought or cold stress. Among the DREB TFs, most of the *DREB2* sub-group is induced by osmotic and heat stresses, and thought to play a key role in the ABA-independent modulation of osmotic stress responses. The up-regulation of *DREB2* genes in heat stress indicates how the response to heat is interconnected with the drought stress response. From the DREB2 sub-group, DREB2A and DREB2B are the TFs best characterized so far, especially in *Arabidopsis* although several homologues have been identified in other species (Matsukura et al. 2010). Both TFs are induced by osmotic and heat-shock stress but not by cold (Nakashima et al. 2000; Sakuma et al. 2006a). Nevertheless, and opposite to the DREB1 TFs, DREB2A did not induce stress tolerance or developmental alterations when over-expressed (Liu et al. 1998). In fact, researchers were able to identify a negative regulatory domain that, when deleted, would cause DREB2A to be constitutively active (Sakuma et al. 2006b) promoting, under non-stress conditions, the expression of many drought-responsive genes such as LEAs as well as heat-shock-responsive genes. This led to the hypothesis of post-translational modifications regulating DREB2A function. Post-translational modifications such as ubiquitination (discussed in section "Chromatin and Epigenomic Regulation") and phosphorylation, affect DREB2A stability and activity. Phosphorylation may be negatively involved, affecting the binding affinity of DREB2A to the DRE motif in pearl millet (Agarwal et al. 2007).

Post-transcriptional and Post-translational Regulation of Drought-Responsive Genes

Expression regulation of key genes involved in adaptation to drought does not solely rely on the action of upstream regulatory proteins since specific changes occurring at the post-transcriptional and post-translational level have been also shown to modulate drought response. MicroRNAs (miRNAs) are a specific class of endogenous small RNAs, typically ~21 nucleotides-long, which can bind to target messenger RNAs through perfect or near-perfect complementarity and lead to the cleavage of target mRNA and repression of translation. MicroRNAs have emerged as important regulators of plant development and stress response and their involvement in drought response has been reported in several species such as Arabidopsis, rice, maize, Medicago and poplar (Liu et al. 2008; Shuai et al. 2013; Trindade et al. 2010; Zhou et al. 2010). The involvement of miRNAs in drought response spans from molecular adaptation pathways, such as osmotic adjustment, antioxidant production and ABA and auxin signalling, to physiological adaptation mechanisms, including stomatal opening, energy maintenance and root/leaf growth and development (reviewed in Ding et al. 2013). Several trancriptomic studies have reported the up- and down-regulation of specific miRNAs in response to drought, pointing out miRNA genes (MIR genes) as targets of upstream signalling pathways. Interestingly, the induction of the miR169 by drought was correlated with the presence of DRE and ABRE motifs in the promoter of the MIR169g gene (Zhao et al. 2007). In addition, miRNA biogenesis may be also regulated by drought through the regulation of ARGONAUTE1 (AGO1), a key factor for miRNA processing. AGO1 is a known target of miR168, which was shown to be differentially regulated under drought stress in Arabidopsis and rice (Liu et al. 2008; Zhou et al. 2010) and could be involved in feedback regulation of miRNA activity in plants under drought stress.

At a post-translational level (similarly to what is observed at transcriptional level) it is possible to identify ABA-dependent and -independent pathways modulating drought stress responses. DREB2A is a known TF that regulates drought stress responses (Sakuma et al. 2006b) in *Arabidopsis* through an ABA-independent manner. Two proteins were described as interacting with DREB2A, DRIP1 and DRIP2, influencing DREB2A stability and thus negatively modulating responses to drought. These proteins were identified as E3 ubiquitin ligases, targeting DREB2A for proteasome degradation (Qin et al. 2008). KEEP ON GOING (KEG) is a RING-ANK (Really Interesting New Gene-Ankyrin) E3 ubiquitin ligase that is involved in ABA signaling (Stone et al. 2006). The *Arabidopsis keg* mutant showed a hypersensitive phenotype to the adverse effects of ABA on development, indicating that KEG could be a negative regulator of ABA-mediated responses. As target for this E3 ligase, ABI5 (ABSCISIC ACID-INSENSITIVE 5), a bZIP TF involved in ABA responses, was found to be more stable in the *keg* mutant. Recently, the *Arabidopsis* KEG E3 ligase was also shown as negatively modulating the stability of two other members of the ABA signaling, ABF1 and ABF3 (ABRE Binding Factors 1 and 3), in an ABA-dependent manner (Chen et al. 2013). In crops, the information regarding this control mechanism of proteins in response to drought is still scarce, but emerging. It is the case of an E3-ubiquitin ligase from wheat, the RING Finger 1 (TdRF1), that was reported to be involved in dehydration response through the modulation of three proteins responsive to either dehydration and/or cold (Guerra et al. 2012).

Chromatin and Epigenomic Regulation

As described in the previous sections, plant response to stress triggers multiple adjustments, at physiological, biochemical and gene expression levels. The reorganization of chromatin structure associated with epigenetic modifications represents an additional layer to modulate gene expression in plant response to environmental challenges (Mcclintock 1984; Arnholdt-Schmitt 2004; Madlung and Comai 2004; Chinnusamy and Zhu 2009; Angers et al. 2010; Kim et al. 2010a; Santos et al. 2011; Han and Wagner 2013).

Epigenetics refers to mitotically or meiotically heritable changes in gene function that cannot be explained by changes in the original DNA sequence (Chinnusamy and Zhu 2009; Jablonka and Raz 2009). A huge network of epigenetic factors has been identified e.g. chromatin remodelers (ChR); DNA methyltransferases (DNMTs); DNA demethylases (DDMs); histone acetyltransferases (HATs); histone deacetylases (HDACs); histone methyltransferases (HMts); histone demethylases (HDMs) but their mechanistic function in modulating gene expression upon specific transcriptional requirements is still not well characterized (Goldberg et al. 2007; Santos et al. 2011). An integrated assessment of epigenomic and transcriptomic datasets under sub-optimal conditions may contribute to better understanding the link between epigenetic regulation and stress tolerance (Li et al. 2008; Zhu 2008; Laird 2010). Epigenetic marks, including DNA methylation, histone modifications, nucleosome positioning and small non-coding RNAs, are not static but instead are transient, meaning that they may reset to initial levels once the stress disappears (Chinnusamy and Zhu 2009). Furthermore, it is plausible to consider that the erasure of epigenetic marks at gametogenesis should not be massive and therefore some epigenetic marks can be meiotically inherited (Saze 2008; Chinnusamy and Zhu 2009; Hauser et al. 2011; Saze 2012). The transgenerational inheritance of epigenetic marks means that the adjustments triggered by stress, may actually be extended to the progeny even in the absence of stress suggesting the plant's capacity to memorize stressful conditions through epigenetic marks, with potential implications in plant breeding (Bruce et al. 2007; Saze 2008; Peng and Zhang 2009; Hauser et al. 2011; Pecinka and Scheid 2012). Whether the plasticity of epigenetic marks and their putative transgenerational inheritance can bring out adaptive advantages regarding better plant performance and yield under stressful environments is far from being understood.

DNA Methylation

The plasticity of DNA methylation has been correlated with plant response to drought. A combined analysis of the methylome and transcriptome dynamics in Arabidopsis seedlings under drought enabled the identification of differentially DNA methylated loci in the promoters of drought responsive genes (Colaneri and Jones 2013). In rice, comparative studies of DNA methylation in drought tolerant *versus* sensitive varieties by using the methylation-sensitive amplified polymorphism method (MSAP) allowed the finding of differential DNA methylation *loci* between contrasting genotypes (Wang et al. 2011a). A similar study in a dry land type of bean (*Macrotyloma uniflorum*) also revealed differential methylation regions in contrasting genotypes (Bhardwaj et al. 2013). Also, in barley genotypes distinguished by their drought tolerance, differential DNA methylation patterns were detected specifically within the promoter and coding regions of a drought stress inducible gene encoding a DNA glycosylase (closely related to cereal DME-family DNA glycosylases, HvDME) suggesting its epigenetic regulation under drought (Kapazoglou et al. 2013). These observations suggest that methylation changes can have a role in the differential drought tolerance of contrasting genotypes. Furthermore, in rice varieties possessing different drought sensitivities, the drought related DNA methylation changes were reported to be transmitted across generations (Zheng et al. 2013). The heritability of methylation changes can have implications both at the transcriptional and the phenotypic levels (Tricker et al. 2013; Niederhuth and Schmitz 2014). For example, in Arabidopsis, increased DNA methylation at specific gene loci (*SPCH* and *FAMA*) involved in the stomatal developmental pathway was associated not only with a decreased gene expression but also with a reduction in the leaf stomatal frequency (Tricker et al. 2012). Continuing this work, the authors were able to find a heritability of those methylation marks across successive generations, which correlated with better tolerance of Arabidopsis plants under drought (Tricker et al. 2013).

Variations in DNA methylation levels under drought have been related to tissue specificity and developmental stages. In rice, global DNA methylation decreased in response to drought, being more pronounced at the tillering stage than at the bolting and heading stages (Wang et al. 2011a). Additionally, contrasting rice varieties at the same developmental stage had less global DNA methylation in roots than in leaf tissues, which may be related to distinctive functions of root tissues in drought sensing (Wang et al. 2011a). Remarkably, in rice 70 % of the drought induced changes in DNA methylation were reported to revert to their original status after a recovery period, while a considerable proportion of about 30 % remained stable (Wang et al. 2011a). In tomato (*Solanum lycopersicum*), the inspection of methylation status in roots after imposing drought treatment focused specifically in the *Asr2*, a protein-coding gene related to alleviation of water scarcity. The analysis of this gene revealed an extensive loss of methylation specifically at the CNN methylation context within the regulatory region (Gonzalez et al. 2013) but the mechanisms underlying chromatin demethylation are still not well understood (Zhang and Zhu 2012).

Globally, these reports illustrate the role of DNA methylation changes in modulating the expression of drought responsive genes with putative implications on drought tolerance, introducing the possibility of using epigenetic-based strategies to improve drought adaptation.

Histone Modifications

The concept of the "histone code" postulates that combinations of histone variants and post-translational modifications via covalent modifications of histone tails by acetylation, methylation, phosphorylation, ubiquitination, biotinylation or sumoylation have a role in modulating chromatin structure which in turn may be involved in physiological responses (Strahl and Allis 2000; Jenuwein and Allis 2001; Berger 2007). Correlations between chromatin and epigenetic factors with physiological and biochemical processes are not well characterized. One clear example is the ATX1 (ARABIDOPSIS TRITHORAX 1) a H3K4 methyltransferase that is involved in drought stress signaling pathways (ABA dependent and ABA independent) and in case of loss of function of atx1, the mutant plants acquire larger stomatal apertures and higher transpiration rates, causing decreased tolerance to drought stress (Ding et al. 2011). Furthermore, in barley a gene encoding a trithorax-like H3K4 methyltransferase, the *HvTX*1 gene showed increased transcription after water scarcity stress in a drought tolerant variety (Papaefthimiou and Tsaftaris 2012). Also, histone deacetylases have been implicated in drought stress signaling, for example the overexpression of the AtHD2C in transgenic *Arabidopsis* plants resulted in ABA insensitivity, which was associated with enhanced drought tolerance (Sridha and Wu 2006).

Dynamic alterations of histone modification profiles have been correlated with the transcriptional regulation of drought responsive genes via chromatin remodeling (Kim et al. 2008; Chinnusamy and Zhu 2009; Granot et al. 2009; van Dijk et al. 2010; Kim et al. 2010a, 2012a; Zong et al. 2013). In plants, specific histone modifications, namely acetylation (e.g. H3K9ac; H3K27ac) and certain methylation (H3K4me3; H3K36me3) marks have been considered to be positively correlated with transcription activation (Chinnusamy and Zhu 2009; van Dijk et al. 2010). The genome-wide analysis of histone methylation at lysine 4 (H3K4) by chromatin immunoprecipitation (ChIP-Seq) has been combined with transcriptomic data in order to investigate the role of H3K4 in the modulation of gene expression under dehydration stress conditions in Arabidopsis (Kim et al. 2008; van Dijk et al. 2010; Kim et al. 2012a) and in rice (Zong et al. 2013). In Arabidopsis, the H3K4 trimethylation (H3K4me3) levels positively correlated with the transcription levels of drought-responsive genes, while H3K4 monomethylation (H3K4me1) showed a negative correlation and the dimethylation (H3K4me2) had no obvious effects (van Dijk et al. 2010). Additionally, the activation of specific drought stress responsive genes in Arabidopsis (*RD*29A, *RD*29B, *RD*20, *RAP*2.4) was associated with changes in lysine modifications at the histone H3 N-tail namely, an enrichment of H3K4me3 and H3K9ac marks in the coding regions during dehydration stress (Kim et al. 2008). This study was extended to the analysis of chromatin status during drought stress recovery and, interestingly, the H3K9 acetylation marks were removed while the H3K4me3 marks were mostly maintained (Kim et al. 2012a). In rice, changes in the H3K4me3 marks in specific dehydrin genes were also positively correlated with their inducible expression under drought (Zong et al. 2013).

Accumulation of Osmolytes/Compatible Solutes

Downstream of these signaling cascades are genes coding for the effectors of response. These can be enzymes that convert substrates into molecules capable of providing some tolerance to the plant. Hydrophytes have the ability to develop structures able to withstand long periods of desiccation. These structures, however, are limited to seeds, pollen or dormant buds, while leaves and other organs usually do not resist water content values below 20–50 %. Some desiccation-tolerant plants have mechanisms that allow them to survive really low levels of relative water content (RWC) in the range of 4–13 % (Otto et al. 2009). One of such mechanisms involves the constitutive expression, or induction, of specific programs to synthesize/accumulate high amounts of osmolytes. Also known as compatible solutes, osmolytes are highly soluble small organic compounds that do not interfere with cellular metabolism, thus being "compatible", even at high concentrations.

The chemical nature of these small molecular weight organic solutes is diverse, including carbohydrates (e.g., fructose, sucrose, trehalose, raffinose), polyols (e.g., mannitol, sorbitol, *myo*-inositol, D-ononitol, D-pinitol) and amino acids (e.g., asparagine, glutamate, proline, serine) as well as amines (e.g., polyamines, glycine-betaine) (Chaves and Oliveira 2004; Krasensky and Jonak 2012). It is generally accepted that the increase in cellular osmolytes functions to stabilize proteins and cellular structures and/or maintain cell turgor by osmotic adjustment. Additionally, it has also been proposed that the accumulation of osmolytes confers protection against the oxidative damage that impairs the normal cell function (Bartels and Sunkar 2005; Yancey 2005; Mittler 2006; Valliyodan and Nguyen 2006).

While the osmoregulatory mechanisms involved are still poorly understood, the increase in osmolytes levels must always be coordinated with both the osmotic and metabolic needs required for plant survival under adverse environmental conditions (Hare et al. 1998).

Trehalose and Other Carbohydrates

The work developed with desiccation-tolerant plants showed the importance of carbohydrate metabolism to effectively confer protection to plants upon dehydration. However, different plants may accumulate different types of sugars in response to drought (Bartels 2005). Trehalose is a non-reducing sugar that functions as carbon source and stress protectant in some desiccation-tolerant plants. This characteristic made the modulation of this sugar concentration an interesting target for developing drought-tolerant plants through genetic engineering (Yeo et al. 2000). However, in the majority of plants, trehalose is present only in trace amounts and abiotic stress periods only slightly modified its content (Garg et al. 2002). The most widely distributed pathway to synthesize trehalose in nature consists of two consecutive enzymatic reactions (Paul et al. 2008). Trehalose-6-phosphate synthase (TPS) produces the intermediate trehalose-6-phosphate (T6P) from UDP-glucose and glucose-6-phosphate followed by a dephosphorylation reaction to trehalose catalyzed by trehalose-6-phosphate phosphatase (TPP). Although the altered expression of genes encoding key-enzymes involved in trehalose biosynthesis, for example trehalose-6-phosphate synthase (TPS), led to improvements in drought, salinity and cold tolerance in species like potato, Arabidopsis or rice (Yeo et al. 2000; Avonce et al. 2004; Ge et al. 2008; Li et al. 2011), trehalose levels accumulated only to trace amounts, excluding a direct role of this osmolyte in mediating osmotic adjustment (Hare et al. 1998; Karim et al. 2007). However, trehalose has been shown to slow down protein aggregation and denaturation (Jain and Roy 2009). Given the importance of water interactions for biological molecules, it is likely that the trehalose protective role *in vivo* is similar to what was described by Dashnau (2007) in *in vitro* reactions. This author showed that extensive intramolecular hydrogen bond networks are able to reduce the direct interaction of monosaccharides with water, thus disrupting water structure to a smaller extent than molecules lacking such networks. Dashnau also demonstrated that in spite of containing the same monosaccharides, the disaccharides trehalose and maltose have a slightly different ability to modify surrounding water structure, probably justifying their different efficacy in protein protection (Dashnau 2007). Furthermore, low concentrations of trehalose were shown to have a protective role of thylacoid membranes against mechanical freeze-thaw damage (Hincha 1989). Hincha suggested that trehalose protection is achieved by bonding to hydrophilic groups at the surface of the thylakoid membrane. Nonetheless, the mechanism by which trehalose acts in protection in concentrations below the osmotically active range is still not completely understood.

Another important class of non-reducing carbohydrates widely distributed in the plant kingdom is the raffinose family oligosaccharides (RFOs) such as raffinose, stachyose and verbascose. RFO biosynthesis begins with the formation of galactinol from *myo*-inositol and UDP-galactose, a reaction catalyzed by galactinol synthase (GolS). Sequential additions of galactose (Gal) units linked to the glucose moiety of sucrose via $\alpha(1\rightarrow6)$ glycosidic linkages leads to the formation of raffinose, stachyose, and verbascose, respectively. It has long been suggested that the primary roles of RFOs in seeds and vegetative tissues are to store and transport carbohydrates, and also to function as osmoprotection molecules against abiotic stresses such as drought (Koster and Leopold 1988; Blackman et al. 1992; Taji et al. 2002; Hannah et al. 2006a). However, their role in drought tolerance is not fully understood. The overexpression of *GolS1* and *GolS2* in *Arabidopsis thaliana* led to a higher accumulation of both galactinol and raffinose, and to a higher tolerance to drought stress (Taji et al. 2002). However, a transgenic approach to knockdown a tonoplast localized sucrose transporter (*PtaSUC4*) in *Populus* resulted in a higher drought tolerance of the transgenic plants with accumulation of sucrose but not of RFOs. The induction of genes related to RFOs was only observed in more drought-sensitive wild-type plants (Frost et al. 2012). Although the role of RFOs in drought tolerance is still far from being fully understood, increasing evidence has been reported for the strong correlation between the accumulation of RFOs, primarily raffinose, stachyose and verbascose, and the development of desiccation tolerance (Koster and Leopold 1988; Blackman et al. 1992; Taji et al. 2002; Peters et al. 2007; António et al. 2008; Gechev et al. 2013).

Starch is the main carbohydrate storage compound for the majority of plants. Drought can induce the degradation of starch into soluble sugars that may function as osmolytes conferring protection to the stress. Recently, work developed in poplar has shown the importance of starch reserves. When starch levels reduce in poplar seedlings following a drought period (due to conversion to soluble sugars), winter frost tolerance is compromised (Galvez et al. 2013). Genes coding for enzymes involved in starch degradation, such as the beta-amylase *BAM1*, have shown an important role in drought response, with mutants showing lower tolerance to osmotic stress (Valerio et al. 2011).

Polyols and Amino Acids

The compatible solutes mannitol and sorbitol (polyols) are also known to be involved in enhancing stress tolerance in response to different abiotic stresses, including drought. This is due to their primary function as macromolecule stabilizers and scavengers of reactive oxygen species (ROS), thus protecting cells from oxidative damage during stress (Stoop et al. 1996).

Similar protective roles have been described for amino acids, proline being the most important compatible solute. Proline is synthesized in plants by two pathways, the glutamate pathway and the ornithine pathway (Kishor et al. 2005; Szabados and Savoure 2010), using Δ-1-pyrroline-5-carboxylate synthetase (P5CS) and ornithine δ-aminotransferase (OAT), respectively, as the two core enzymes involved in proline biosynthesis (Verbruggen and Hermans 2008). Research in proline metabolism has revealed that proline biosynthesis and catabolism appears to be coordinately regulated in response to water stress. Proline synthesis is activated during stress conditions and degradation is enhanced during stress recovery (Yoshiba et al. 1997; Mattioli et al. 2009; An et al. 2013). It is expected that the overexpression of biosynthetic proline enzymes may lead to the accumulation of this compatible solute and increased tolerance against abiotic stress. Several studies have tested this, and the overexpression of the enzyme P5CS in transgenic plants of tobacco and petunia under drought led to the accumulation of the cellular levels of proline which acted as an osmoprotectant, and therefore, induced drought tolerance (Kishor et al. 1995; Yamada et al. 2005).

Polyamines

Interest has also been growing in the involvement of polyamines (Pas—highly charged small aliphatic polycations derived from amino acids) in conferring protection from a variety of abiotic stresses, including drought (Yang et al. 2007; Bae et al. 2008; Groppa and Benavides 2008; Gill and Tuteja 2010; Alcazar et al. 2011; Bitrian et al. 2012). Polyamines like putrescine (diamine), spermidine (triamine) and spermine (tetramine), are ubiquitous in nature. In plants, putrescine (Put) is derived from either arginine or ornithine by two different pathways, the arginine decarboxylase pathway (ADC pathway) or the ornithine decarboxylase pathway (ODC pathway). Spermidine (Spd) is produced from Put by the addition of an aminopropyl moiety from *S*-adenosylmethionine decarboxylase (SAMDC) in the presence of the enzyme spermidine synthase. Spermine (Spm) is produced from the reaction of Spd with SAMDC, a reaction catalyzed by the enzyme spermine synthase (Carbonell and Blazquez 2009).

The modulation of polyamine levels in plants by the overexpression and/or downregulation of the core enzymes involved in polyamine biosynthesis (ADC, ODC and SAMDC) have long proven to be a valuable tool for investigating their role in plant protection from abiotic stresses (Bhatnagar et al. 2002; Kakkar and Sawhney 2002). An interesting study by Capell and co-workers (2004) showed that transgenic rice plants expressing the *Datura stramonium adc* gene produced much higher levels of putrescine than wild-type plants when exposed to drought stress. Spermidine and spermine biosynthesis was also enhanced, and these plants were shown to have increased drought tolerance.

In a transcriptomics study of 21 rice cultivars including *indica* and *japonica* subspecies, Do et al. (2013) studied the regulation of polyamine biosynthesis under long-term drought-stress. The expression of 21 genes encoding enzymes involved in polyamine biosynthesis was analyzed by RT-qPCR, and the genomic analysis revealed that 11 of these genes were located in drought-related quantitative trait *loci* (QTL) regions. Metabolomics using GC-MS metabolite profiling revealed drought-induced changes in the levels of ornithine and arginine, substrates in the polyamine biosynthetic pathway, and an increase in the levels of spermine. In this study, the combination of gene expression and metabolite data was consistent with a coordinated adjustment of polyamine biosynthesis for the accumulation of spermine under drought conditions, which is in agreement with a role of polyamine metabolism in protecting plants from drought stress (Do et al. 2013).

Although the accumulation of small molecular weight organic solutes (osmolytes/compatible solutes) can help plants to cope with adverse conditions, including drought, the total osmolyte pool accumulating in plants upon drought is far from being completely identified and functionally characterized. This will require not only a better understanding of the processes involved in the production and transport of compatible solutes, but also a better elucidation of the molecular events responsible for plant stress tolerance.

Drought-Affected Developmental Traits

Regulation of Root Traits

Although of recognized importance, root traits have been overlooked by researchers (Den Herder et al. 2010). Root traits are normally controlled by multiple genes and are difficult to phenotype. This may justify that there are almost 400 QTL described as related to drought response in rice shoots, while for roots only 139 are known (Comas et al. 2013). When targeting drought tolerance at root level researchers have to look for traits that may be affected by genotype, environment and management conditions. Potential benefits verified under drought in greenhouse conditions, may be lost in field conditions (Watt et al. 2013). In most cases, the main root traits related to drought avoidance/tolerance are smaller root diameter, increased number and length of root hairs as well as some mechanical aspects of xylem vessels (specific pit anatomy can reduce cavitation under water scarcity). Also, a deeper rooting system (related to *DRO1* expression in rice), higher root biomass with changes in allometry (root to shoot ratio), capacity to maintain root growth under mild and moderate stress and the expression of genes coding for aquaporins are useful traits to improve water uptake (Comas et al. 2013).

For some of these traits there were already QTLs described and identified. The majority of the QTLs identified in species such as rice, maize and wheat are involved in root length, root biomass, root number and root angle. As an example in rice, from 119 QTLs identified as involved in drought responses, the large majority is associated with maximum root length (Courtois et al. 2009). This is not surprising as traits like deeper rooting and root biomass are the easiest to phenotype and primordial drivers for drought avoidance. Recently, several advances in root phenotyping were achieved which may help to understand plant performance under drought and in designing new strategies to improve crop productivity (revised by Selvaraj et al. 2013). However, even though root phenotyping imaging techniques have evolved, they are performed in greenhouse conditions (pots or similar) and normally under specific media to facilitate imaging which can artificially lead to differences in root development. Nevertheless, and throughout the years, researchers have identified, in crops like rice and maize, positive effects of different QTLs and specific genes in the regulation of root traits. It is the case of the rice qtl12.1 for which an allele accounting for a 7 % increase in water uptake explained a large effect on yield improvement (Bernier et al. 2009). Positive effects were also observed with the over-expression of the transcription factor *OsNAC5*, which led to increased drought tolerance, in this case associated to increased root diameter (Jeong et al. 2013), while the maize QTL Root-ABA1d was found to directly affect root architecture and biomass, and indirectly affect leaf ABA levels (Giuliani et al. 2005).

An example on how root traits can affect plant tolerance to water deficit, was given by Uga et al. (2013) who described in an upland rice cultivar (Kinandang Patong, KP), the action of DRO1 gene (*DEEPER ROOTING 1*) recently mapped on chromosome 9, in a region responsible for deeper rooting. When this gene was transformed into the genetic background of a drought-sensitive rice cultivar with a small root system (IR64), the transgenic plant had roots with more than double the length of those of IR64, while showing improved yield in both drought and well-watered conditions (Uga et al. 2013). The roots of the transgenic OX-DRO1 had a faster gravitropic response than IR64 suggesting auxin involvement. In fact, the *DRO1* promoter contains *cis*-elements that allow the binding of auxin responsive factors (ARFs, known repressors of gene expression). Interestingly, other monocots such as maize, which have longer root systems, also have *DRO1* homologues.

Despite the advances on the comprehension of the function of several genes and QTLs involved in root traits under drought stress, still a large effort is needed to integrate the different aspects of genotype × environment × management to reach reasonably improved crop yields under water limiting conditions.

Modulation of Flower Transition

As already mentioned, in a drought escape strategy, drought stress can accelerate the transition to flowering and thus shorten the plant life cycle allowing it to escape the unfavourable conditions. In Arabidopsis plants exposed to drought, the transcription of the florigen genes *FLOWERING LOCUS T* (*FT*), and *TWIN SISTER OF FT* (*TSF*) is up-regulated in an ABA- and photoperiod (long days)-dependent manner (Riboni et al. 2013). The flowering promoting gene *GIGANTEA* (*GI*) enables the drought escape response via ABA-dependent activation of *FT*, *TSF* and *SUPPRESSOR OF OVEREXPRESSION OF CONSTANS* (*SOC1*), while *SOC1* activation contributes to the *TSF* induction. According to Riboni et al. (2013) in the absence of the floral repressor SHORT VEGETATIVE PHASE or in *GI*-overexpressing plants the drought-escape response was recovered under short days.

Recently, the rice gene coding for GRAIN NUMBER, PLANT HEIGHT, AND HEADING DATE7 (Ghd7) was found to be strongly inhibited by drought, ABA and JA (Weng et al. 2014). The authors related this response with the plant's ability to quickly end the life cycle in adverse conditions in order to escape or avoid stresses. Weng et al. (2014) further suggested that Ghd7 integrates the dynamic environmental inputs with phase transition, architecture regulation, and stress response, to maximize rice reproductive success. Moreover, their results showed that *Ghd7* regulates stress-related genes and ROS homeostasis genes.

MicroRNAs have also been described as involved in stress-induced flowering. It is the case of the microRNA family miR169 that is up-regulated by abiotic stress (cold, drought, salinity). Specifically the overexpression of *miR169d* induces early flowering in Arabidopsis, soybean and maize, and the overexpression of one of the *miR169d* targets, the transcription factor *AtNF-YA2* (specially a *miR169d* resistant version), was able to delay flowering (Xu et al. 2014). *AtNF-YA2* normally promotes flowering by inducing FLOWERING LOCUS C (*FLC*), which in turn induces *FT* and LEAFY (*LFY*) (Xu et al. 2014).

Regulation of Stomatal Development and Aperture

It was suggested that stomatal closure is the earliest response to drought and the dominant limitation to photosynthesis at mild to moderate stress and that, concomitantly, there is a progressive reduction in the biochemical processes that become dominant at severe stress leading to reduced photosynthetic CO_2 assimilation (Flexas and Medrano 2002). Considering the relevance of stomata in drought stress response, we review below important aspects of the molecular control of stomata development/aperture and drought response.

Drought and the Regulation of Stomatal Development

Although the molecular mechanisms regulating stomatal development have been well characterized (Pillitteri and Torii 2012), little is known about how drought can modulate this process. Plants are known to regulate their number of stomata in order to optimize gas exchange in response to environmental cues. Increased carbon dioxide in the atmosphere, increased temperatures and changes in light quality all affect stomatal density (Casson and Gray 2008; Casson et al. 2009). It has also been shown that water availability (Wu et al. 2009) as well as exogenous applied ABA (Franks and Farquhar 2001) can modulate stomatal development. The ABA-overly sensitive mutant *abo1* shows a drought-resistant phenotype. The *abo1* mutation enhances ABA-induced stomatal closing and also influences the development of guard cells, resulting in stomata reduction to half the number observed in the wild type (Chen et al. 2006). Recently, it was proposed that ABA limits stomatal initiation (Tanaka et al. 2013). This suggests that ABA signaling must be involved in the modulation of stomatal development in response to drought. It is predictable that, when perceived by plant cells, the water deficit signal is transduced through an ABA-dependent signaling pathway that will converge on the proteins involved in stomatal development (e.g. SPCH, MUTE). Brassinosteroids, which are also known to be involved in drought tolerance, inhibit stomatal development by GSK3-mediated inhibition of a MAPK pathway (Kim et al. 2012b; Khan et al. 2013). The fact that

AtMPK3 and AtMPK6 are directly involved in the signaling cascade that regulates stomatal development and are both responsive to different stresses (Beckers et al. 2009; Xie et al. 2009), suggests that SPCH activity may be directly affected by adverse environmental conditions thereby enabling the plant to modify stomatal development in response to abiotic stress.

It was shown that when subjected to transient mild drought, meristemoids pause divisions but can resume them when re-watered (Skirycz et al. 2011). Recently, it was also shown that drought induced an alteration in stomatal development in two genotypes of *Populus balsamifera* (Hamanishi et al. 2012). Leaves that developed under water-deficit conditions had lower stomatal indices than leaves that developed under well-watered conditions. In addition, the homologues of *STOMAGEN*, *ERECTA*, *STOMATA DENSITY AND DISTRIBUTION 1* (*SDD1*), and *FAMA* showed a variable transcript abundance pattern congruent with their role in the modulation of stomatal development in response to drought. In Arabidopsis, the *STOMAGEN* expression seems to be differentially regulated by ABA, drought, cold, high salinity, and heat treatments (https://www.genevestigator.com).

Some players regulating the integration of environmental cues and stomatal development have already been identified. HIGH CARBON DIOXIDE protein regulates stomatal development in response to CO_2 and the PHYTOCROME INTERACTING FACTOR 4 regulates stomatal development in response to light (Gray et al. 2000). Nevertheless, the integration point of drought signaling into the stomatal development pathway remains unknown. Identification and analysis of mutants under variable conditions will provide insight into this complex system.

Drought and the Regulation of Stomatal Aperture

Stomata control CO_2 influx for photosynthesis and water loss through transpiration. In response to drought, ABA level increases leading to stomata closure and consequent reduction in water loss (Seki et al. 2002; Rabbani et al. 2003; Christmann et al. 2007). ABA-induced stomatal closure requires the coordinate control of guard-cell turgor, cytoskeleton organization, membrane trafficking and gene expression (Hetherington 2001) and is mediated by a complex guard-cell signaling network of kinases/phosphatases, secondary messengers, and ion channel regulators (Kim et al. 2010b).

Under control conditions, the low levels of ABA leave the PP2C protein ABI1 free to inhibit SnRK2 kinase activity (see section "ABA-Dependent Response") (Fig. 2). Under water deficit conditions (high ABA levels), the ABA-receptor complexes interact with PP2C to inhibit its activity and this will allow the SnRK2 protein kinase SnRK2.6/OST1 (OPEN STOMATA 1) to function as positive regulator of ABA-induced stomatal closure (Mustilli et al. 2002). Once active, SnRK2.6/OST1 activates the S-type anion channel SLAC1 (SLOW ANION CHANNEL ASSOCIATED 1) (Geiger et al. 2009; Lee et al. 2009) and inhibits the K^+ inward channel KAT1 by phosphorylation (Sato et al. 2009). SLAC1 is also regulated by calcium-dependent protein kinases (CDPK) (Geiger et al. 2010) and it was recently shown that there are two alternate signaling cores including either CPK6 or SnRK2.6/OST1 (Brandt et al. 2012). The guard-cell outward rectifying K^+-channel GORK is up-regulated by drought and its regulation is mediated by ABI1 and ABI2 (Becker et al. 2003); however, whether SnRK2.6/OST1 (or other protein downstream PP2C) is also involved in GORK regulation is not yet known. SnRK2.6/OST1 also activates the AtRBOHF (RESPIRATORY BURST OXIDASE HOMOLOGUE F), a plasma membrane localized NADPH oxidase that generates H_2O_2 (Sirichandra et al. 2009). H_2O_2 increase mediates stomatal closure through activation of Ca^{2+} channels (Pei et al. 2000). ABA-induced stomatal closure is also mediated by the vacuolar potassium channel TPK (TWO PORE K^+ CHANNEL1) (Gobert et al. 2007) and the ABC transporter AtMRP5 (MULTIDRUG RESISTANT PROTEIN 5), which is a high affinity IP6 transporter (Nagy et al. 2009). GPA1, mentioned above as involved in transpiration efficiency through regulation of stomata development (Nilson and Assmann 2010), also regulates potassium and anion channels in guard-cells (Wang et al. 2001).

The control of stomatal aperture by ABA signaling involves the regulation of many genes. Several transcription factors (TFs) from the subclass MYB-R2R3 have been shown to be involved in stomatal movements in response to environmental conditions. *AtMYB60* was reported to be specifically expressed in guard-cells, and its transcription down-regulated by ABA and water deficit, concomitantly with stomatal closure (Cominelli et al. 2005). The *atmyb60-1* null mutant shows a constitutive reduction in stomatal opening and decreased wilting under drought. In addition, many genes altered in *atmyb60-1* (e.g. *Aquaporin*, *ERD10*, *ERD13*, and *ERF*) are known to be involved in drought response (Cominelli et al. 2005), indicating that AtMYB60 has a broad role regulating this stress response. SCAP1 was shown to directly bind to the MYB60 promoter region, being one of the Dof factors responsible for the regulation of guard-cell specific gene expression (Negi et al. 2013). *AtMYB61* was also shown to mediate light-induced increased stomatal aperture (Liang et al. 2005), but its transcription level was not altered in plants treated with ABA or subjected to drought, indicating that AtMYB61 functions in a mechanism parallel to that responsible for stomata closure in response to drought. Nevertheless, a post-transcriptional/translational regulation of AtMYB61 cannot be ruled out. Another MYB TF involved in stomatal closure in response to drought stress is AtMYB44 (Jung et al. 2008). *AtMYB44* is expressed in the vasculature and leaf epidermal guard-cells and its transcription is rapidly activated (within 30 min) upon dehydration and ABA treatment. In addition, transgenic Arabidopsis plants over-expressing AtMYB44 showed higher sensitivity to ABA, more rapid ABA-induced stomatal closure than wild type, reduced rate of water loss, and consequently an enhanced abiotic stress tolerance. Other TFs involved in stomatal aperture regulation in response to drought are SNAC1 (Hu et al. 2006) and DROUGHT AND SALT TOLERANCE (DST) (Huang et al. 2009). Over-expression of SNAC1 was shown to improve drought tolerance in rice by increasing ABA sensitivity and stomatal closure, leading to decreased water lost (Hu et al. 2006). DST is a zinc finger that regulates drought and salt tolerance by direct modulation of genes related to H_2O_2 homeostasis. The DST-mediated pathway operates in an ABA-independent manner. Besides stomatal closure, DST is also involved in the regulation of stomatal density, indicating a putative function in the drought control of stomata development.

Regulation of Photosynthesis

Plants can cope with water deficit conditions by increasing the protection of the photosynthetic system from oxidative damage (e.g., reducing the activity of photosynthetic enzymes or affecting electron transport). Transcript profiling studies have shown that most of the genes involved in photosynthesis (PSI, PSII, Calvin cycle, LHCI, and LHCII) are down-regulated in plants subjected to water stress. Modulation of gene expression suggests that decrease of photosynthetic capacity under water stress conditions is not only due to damage of the photosynthetic apparatus, but may be a regulatory response. The down-regulation of several photosynthesis-related genes under water deficit have been observed in different plant species. In barley, among the different categories of genes down-regulated under water deficit, the most drastic reduction was observed for photosynthesis-related genes (Ozturk et al. 2002). Chlorophyll a/b-binding protein precursor, Rubisco small subunit, and Ribulose-bisphosphate carboxylase activase are among the most down-regulated genes. Another study in barley showed that the transcripts strongly down-regulated in response to water deficit are related to photosynthesis, photorespiration, and metabolism of amino acids and carbohydrates (Talame et al. 2007). When rice was subjected to water deficit, photosynthesis was the only process to have a notably higher number of down-regulated genes as compared with the up-regulated ones (Hazen et al. 2005). In *Populus*, photosynthesis was the most over-represented biological process in functional enrichment analysis of the down-regulated genes (Tang et al. 2013). Several photosynthesis-related genes, such as chlorophyll a/b-binding protein CP24, photosystem I reaction center subunit V, protochlorophyllide reductase A, peptidyl-prolyl *cis-trans* isomerase, and others functioning in the process of photosynthesis, were uniquely down-regulated in rice leaves subjected to water deficit (Wang et al. 2011b). In wheat, the expression levels of most genes involved in carbon fixation were also reduced in the leaves during prolonged water deficit. In addition, a correlation analysis indicated that the genes encoding key enzymes of the Calvin cycle were coordinately down-regulated (Xue et al. 2008).

It has been observed that water deficit affects photosynthesis efficiency and the expression of associated genes in a dose-dependent manner. Analysis of the publicly available expression profiling data under drought stress showed a non-significant effect (both qualitatively and quantitatively) of mild drought on the expression of photosynthetic genes (Chaves et al. 2009). Microarray analysis of Arabidopsis plants subjected to water deficit revealed that many photosynthesis genes were significantly repressed under severe water stress in contrast to the subtle effect of the moderate stress (Harb et al. 2010). In *Populus*, drought only slightly impacted photosynthesis parameters under mild and moderate drought treatment in contrast to photosynthesis inhibition under severe drought (Tang et al. 2013). Under moderate drought, various genes related to PSI and PSII reaction center, Calvin cycle and chlorophyll A/B binding proteins were repressed under water deficit conditions, but the fold change was relatively low.

Although most photosynthesis-related genes are down-regulated under water deficit conditions, some photosynthesis-related genes can be induced under stress (Ji et al. 2012; Tang et al. 2013). Tang and collaborators (2013) suggested that up-regulation of few photosynthesis-related genes may contribute to maintain photosynthesis under water deficit. In the drought tolerant rice variety IRAT109, it was observed that while Rubisco was down-regulated, Rubisco activase was up-regulated (Ji et al. 2012). This up-regulation might alleviate the damage caused by the lower level of Rubisco under drought.

The way photosynthesis performance is regulated under adverse environmental conditions can make plants more or less tolerant to stress. It was shown that the magnitude of the photosynthesis gene repression is positively correlated with the freezing tolerance of different Arabidopsis accessions (Hannah et al. 2006b). Genes related to photosynthesis, particularly Calvin cycle related genes, were down-regulated mainly in the drought-tolerant genotypes as compared to susceptible genotypes under drought conditions (Degenkolbe et al. 2009; Hayano-Kanashiro et al. 2009). On the other hand, upon recovery (irrigation) the expression pattern was reversed, with an increase in differential expression of photosynthesis-related genes, higher in tolerant genotypes than in susceptible ones (Degenkolbe et al. 2009). Down-regulation of photosynthetic genes in the tolerant cultivars may indicate an adaptive response to prevent photodamage in periods of reduced CO_2 availability in the mesophyll, when stomata are closed due to water shortage (Degenkolbe et al. 2009). Under water deficit conditions, the tolerant cultivars had a higher photosynthetic capacity and produced more biomass than the sensitive ones. In addition, the gene expression of the photosystem II protein D2, photosystem II 44 kDa protein, two chlorophyll a/b binding proteins, the photosystem I reaction center subunits III and IX, ribulose bisphosphate carboxylase small subunit C and the alpha and beta chains of cytochrome b559, was specifically down-regulated by water deficit in the tolerant rice cultivars (Degenkolbe et al. 2009). Regulation of the photosynthesis-related genes plays a critical role in drought tolerance.

Energy Imbalance

The stress sensed by the plant results in impaired carbon assimilation and/or respiration, which is known to trigger the activation of protein kinases of the SnRK1 [Snf1 Sucrose non-fermenting 1-related protein kinases 1] family. These enzymes are able to modify the metabolic and transcriptional program in order to restore homeostasis (Rodrigues et al. 2013). As global regulators of carbon metabolism,

SnRKs have 38 members in plants, distributed by three sub-groups (SnRK1, SnRK2 and SnRK3) linking metabolic and stress signaling (Coello et al. 2011).

In spite of the difficulty in studying this system, the SnRK path has been gradually uncovered, giving sugars a particular relevance in signaling and regulation (Rolland et al. 2006), beyond their well known osmotic role (see sections "ABA-Dependent Response" and "Accumulation of Osmolytes/Compatible Solutes"). As highlighted by Baena-Gonzalez et al. (2010), failure to set up an initial 'emergency' response may end up in nutrient deprivation and irreversible senescence and cell death.

Although not specific of drought, energy deficiency resulting from the stress will necessarily impact photosynthesis and/or respiration. Therefore, the role of SnRKs in the metabolic/stress signaling interface makes them potential targets for crop adaptation to drought environments.

Challenges and Opportunities for Plant Improvement

Plant stress responses are dynamically interconnected with different levels of regulation involving fine adjustments of metabolism but also of gene expression, which are two processes consistent with the known complexity of the functional network that controls stress tolerance. Therefore, when studying stress-related plant responses, a wide range of parameters must be considered, ranging from the morphophysiology to gene expression, metabolism and epigenomics. There is still a long way before direct correlations can be established between epigenetic marks and specific traits of interest, such as a reasonable performance under stress. For instance, it remains to be understood how distinct epigenetic marks act in a coordinated manner to generate genomic responses to sub-optimal environmental challenges. A targeted manipulation of epigenetic marks can be envisaged as a way to modulate gene expression of key drought tolerance-related genes. Another crucial research avenue is to better understand the role of plant epigenetic plasticity in the communication of environmental conditions to the offspring, as a way to confer advantages in coping with adversities (Witzany 2006; Bruce et al. 2007; Hauser et al. 2011; Slaughter et al. 2012). In this context, further experimentation is needed to address the discrimination of transient/reversible epigenetic marks from those that are retained and transmitted to the next generation.

To date, due to the complex nature of plant responses, there are only a small number of genes that have demonstrated a beneficial adaptation under water limiting conditions with confirmed yield increases and non-negative effects when water availability is not a problem. This is the case of genes involved in the increase of threalose accumulation (e.g. *tps* and *tpp* genes, both involved in threalose biosynthesis), other osmolytes (raffinose family oligossacharides, *GolS* or proline, *P5CS*) or even polyamines (*odc*). Other genes have also shown a positive effect in the response to water limiting conditions, namely transcription regulators [e.g. SNAC1 (Hu et al. 2006; Liu et al. 2014)], ABA-receptors [e.g. PYL5 (Kim et al. 2014)] and developmental regulators such as regulators of root-specific traits [e.g. DRO1 (Uga et al. 2013)].

Large-scale transcriptomics, proteomics and metabolomics, combined with studies of expression regulation, biochemistry, physiology and phenotyping under stress are raising the data necessary for deeper holistic approaches. In this endeavour, good phenotyping in field conditions and systems biology strategies may help to raise comprehensive models of plant behaviour. This knowledge is needed to identify new strategies to improve crop productivity and develop drought-tolerant plants, while also helping to better elucidate how plants coordinate growth and development in a constantly changing environment. Moreover, given that in nature it is rare that one stress occurs alone, understanding how plants cope with combined stresses is crucial to advance breeding programs for improved crops.

Acknowledgements Funding of the Research unit GREEN-it "Bioresources for Sustainability" (UID/Multi/04551/2013) is also gratefully acknowledged.

References

Agarwal P, Agarwal PK, Nair S, Sopory SK, Reddy MK (2007) Stress-inducible DREB2A transcription factor from *Pennisetum glaucum* is a phosphoprotein and its phosphorylation negatively regulates its DNA-binding activity. Mol Genet Genomics 277:189–198

Alcazar R, Bitrian M, Bartels D, Koncz C, Altabella T, Tiburcio AF (2011) Polyamine metabolic canalization in response to drought stress in *Arabidopsis* and the resurrection plant *Craterostigma plantagineum*. Plant Signal Behav 6:243–250

Almeida T, Menendez E, Capote T, Ribeiro T, Santos C, Goncalves S (2013) Molecular characterization of *Quercus suber MYB1*, a transcription factor up-regulated in cork tissues. J Plant Physiol 170:172–178

Almeida-Rodriguez AM, Cooke JE, Yeh F, Zwiazek JJ (2010) Functional characterization of drought-responsive aquaporins in *Populus balsamifera* and *Populus simoniixbalsamifera* clones with different drought resistance strategies. Physiol Plant 140:321–333

Alvarez S, Marsh EL, Schroeder SG, Schachtman DP (2008) Metabolomic and proteomic changes in the xylem sap of maize under drought. Plant Cell Environ 31:325–340

An YY, Zhang MX, Liu GB, Han RL, Liang ZS (2013) Proline accumulation in leaves of *Periploca sepium* via both biosynthesis up-regulation and transport during recovery from severe drought. PLoS One 8, e69942

Angers B, Castonguay E, Massicotte R (2010) Environmentally induced phenotypes and DNA methylation: how to deal with unpredictable conditions until the next generation and after. Mol Ecol 19:1283–1295

Antonio C, Pinheiro C, Chaves MM, Ricardo CP, Ortuno MF, Thomas-Oates J (2008) Analysis of carbohydrates in *Lupinus albus* stems on imposition of water deficit, using porous graphitic carbon liquid chromatography-electrospray ionization mass spectrometry (vol 1187, pg 111, 2008). J Chromatogr A 1201:132

Arnholdt-Schmitt B (2004) Stress-induced cell reprogramming. A role for global genome regulation? Plant Physiol 136:2579–2586

Avonce N, Leyman B, Mascorro-Gallardo JO, Van Dijck P, Thevelein JM, Iturriaga G (2004) The Arabidopsis trehalose-6-P synthase AtTPS1 gene is a regulator of glucose, abscisic acid, and stress signaling. Plant Physiol 136:3649–3659

Bae H, Kim SH, Kim MS, Sicher RC, Lary D, Strem MD et al (2008) The drought response of *Theobroma cacao* (cacao) and the regulation of genes involved in polyamine biosynthesis by drought and other stresses. Plant Physiol Biochem 46:174–188

Baena-Gonzalez E (2010) Energy signaling in the regulation of gene expression during stress. Mol Plant 3:300–313
Baker SS, Wilhelm KS, Thomashow MF (1994) The 5′-region of *Arabidopsis thaliana* cor15a has cis-acting elements that confer cold-, drought- and ABA-regulated gene expression. Plant Mol Biol 24:701–713
Bartels D (2005) Desiccation tolerance studied in the resurrection plant *Craterostigma plantagineum*. Integr Comp Biol 45:696–701
Bartels D, Sunkar R (2005) Drought and salt tolerance in plants. Crit Rev Plant Sci 24:23–58
Bartoli CG, Gomez F, Gergoff G, Guiamet JJ, Puntarulo S (2005) Up-regulation of the mitochondrial alternative oxidase pathway enhances photosynthetic electron transport under drought conditions. J Exp Bot 56:1269–1276
Baxter A, Mittler R, Suzuki N (2013) ROS as key players in plant stress signalling. J Exp Bot
Beadle NCW (1966) Soil phosphate and its role in molding segments of Australian flora and vegetation with special reference to xeromorphy and sclerophylly. Ecology 47:992–1007
Becker D, Hoth S, Ache P, Wenkel S, Roelfsema MR, Meyerhoff O et al (2003) Regulation of the ABA-sensitive Arabidopsis potassium channel gene GORK in response to water stress. FEBS Lett 554:119–126
Beckers GJM, Jaskiewicz M, Liu YD, Underwood WR, He SY, Zhang SQ et al (2009) Mitogen-activated protein kinases 3 and 6 are required for full priming of stress responses in *Arabidopsis thaliana*. Plant Cell 21:944–953
Berger SL (2007) The complex language of chromatin regulation during transcription. Nature 447:407–412
Bernier J, Serraj R, Kumar A, Venuprasad R, Impa S, Gowda RPV et al (2009) The large-effect drought-resistance QTL qtl12.1 increases water uptake in upland rice. Field Crop Res 110:139–146
Bhardwaj J, Mahajan M, Yadav SK (2013) Comparative analysis of DNA methylation polymorphism in drought sensitive (HPKC2) and tolerant (HPK4) genotypes of horse gram (*Macrotyloma uniflorum*). Biochem Genet 51:493–502
Bhatnagar P, Minocha R, Minocha SC (2002) Genetic manipulation of the metabolism of polyamines in poplar cells. The regulation of putrescine catabolism. Plant Physiol 128:1455–1469
Bitrian M, Zarza X, Altabella T, Tiburcio AF, Alcazar R (2012) Polyamines under abiotic stress: metabolic crossroads and hormonal crosstalks in plants. Metabolites 2:516–528
Blackman SA, Obendorf RL, Leopold AC (1992) Maturation proteins and sugars in desiccation tolerance of developing soybean seeds. Plant Physiol 100:225–230
Bogeat-Triboulot MB, Brosche M, Renaut J, Jouve L, Le Thiec D, Fayyaz P et al (2007) Gradual soil water depletion results in reversible changes of gene expression, protein profiles, ecophysiology, and growth performance in *Populus euphratica*, a poplar growing in arid regions. Plant Physiol 143:876–892
Bosu PPP, Wagner MRM (2007) Effects of induced water stress on leaf trichome density and foliar nutrients of three elm (*Ulmus*) species: implications for resistance to the elm leaf beetle. Environ Entomol 36:595–601
Boudsocq M, Barbier-Brygoo H, Lauriere C (2004) Identification of nine sucrose nonfermenting 1-related protein kinases 2 activated by hyperosmotic and saline stresses in *Arabidopsis thaliana*. J Biol Chem 279:41758–41766
Brandt B, Brodsky DE, Xue SW, Negi J, Iba K, Kangasjarvi J et al (2012) Reconstitution of abscisic acid activation of SLAC1 anion channel by CPK6 and OST1 kinases and branched ABI1 PP2C phosphatase action. Proc Natl Acad Sci U S A 109:10593–10598
Brodersen CR, McElrone AJ (2013) Maintenance of xylem network transport capacity: a review of embolism repair in vascular plants. Front Plant Sci 4:108
Brodersen CR, McElrone AJ, Choat B, Lee EF, Shackel KA, Matthews MA (2013) In vivo visualizations of drought-induced embolism spread in *Vitis vinifera*. Plant Physiol 161:1820–1829
Bruce TJA, Matthes MC, Napier JA, Pickett JA (2007) Stressful "memories" of plants: evidence and possible mechanisms. Plant Sci 173:603–608

Cao K-F, Yang S-J, Zhang Y-J, Brodribb TJ (2012) The maximum height of grasses is determined by roots. Ecol Lett 15:666–672
Capell T, Bassie L, Christou P (2004) Modulation of the polyamine biosynthetic pathway in transgenic rice confers tolerance to drought stress. Proc Natl Acad Sci U S A 101:9909–9914
Carbonell J, Blazquez MA (2009) Regulatory mechanisms of polyamine biosynthesis in plants. Genes Genomics 31:107–118
Casson S, Gray JE (2008) Influence of environmental factors on stomatal development. New Phytol 178:9–23
Casson SA, Franklin KA, Gray JE, Grierson CS, Whitelam GC, Hetherington AM (2009) phytochrome B and PIF4 regulate stomatal development in response to light quantity. Curr Biol 19:229–234
Chaves MM, Oliveira MM (2004) Mechanisms underlying plant resilience to water deficits: prospects for water-saving agriculture. J Exp Bot 55:2365–2384
Chaves MM, Maroco JP, Pereira JS (2003) Understanding plant responses to drought – from genes to the whole plant. Funct Plant Biol 30:239
Chaves MM, Flexas J, Pinheiro C (2009) Photosynthesis under drought and salt stress: regulation mechanisms from whole plant to cell. Ann Bot 103:551–560
Chaves MM, Costa JM, Saibo NJM (2011) Recent advances in photosynthesis under drought and salinity. Plant responses to drought and salinity stress: developments in a post-genomic era. Adv Bot Res 57:49–104
Chen ZZ, Zhang HR, Jablonowski D, Zhou XF, Ren XZ, Hong XH et al (2006) Mutations in ABO1/ELO2, a subunit of holo-elongator, increase abscisic acid sensitivity and drought tolerance in *Arabidopsis thaliana*. Mol Cell Biol 26:6902–6912
Chen YT, Liu HX, Stone S, Callis J (2013) ABA and the ubiquitin E3 ligase KEEP ON GOING affect proteolysis of the *Arabidopsis thaliana* transcription factors ABF1 and ABF3. Plant J 75:965–976
Chinnusamy V, Zhu JK (2009) Epigenetic regulation of stress responses in plants. Curr Opin Plant Biol 12:133–139
Choat B, Jansen S, Brodribb TJ, Cochard H, Delzon S, Bhaskar R et al (2012) Global convergence in the vulnerability of forests to drought. Nature 491:752–755
Christmann A, Hoffmann T, Teplova I, Grill E, Muller A (2005) Generation of active pools of abscisic acid revealed by in vivo imaging of water-stressed Arabidopsis. Plant Physiol 137:209–219
Christmann A, Weiler EW, Steudle E, Grill E (2007) A hydraulic signal in root-to-shoot signalling of water shortage. Plant J 52:167–174
Coello P, Hey SJ, Halford NG (2011) The sucrose non-fermenting-1-related (SnRK) family of protein kinases: potential for manipulation to improve stress tolerance and increase yield. J Exp Bot 62:883–893
Colaneri AC, Jones AM (2013) Genome-wide quantitative identification of DNA differentially methylated sites in Arabidopsis seedlings growing at different water potential. PLoS One 8, e59878
Comas LH, Becker SR, Cruz VM, Byrne PF, Dierig DA (2013) Root traits contributing to plant productivity under drought. Front Plant Sci 4:442
Cominelli E, Galbiati M, Vavasseur A, Conti L, Sala T, Vuylsteke M et al (2005) A guard-cell specific MYB transcription factor regulates stomatal movements and plant drought tolerance. Curr Biol 15:1196–1200
Courtois B, Ahmadi N, Khowaja F, Price AH, Rami JF, Frouin J et al (2009) Rice root genetic architecture: meta-analysis from a drought QTL database. Rice 2:115–128
Cramer GR, Ergul A, Grimplet J, Tillett RL, Tattersall EAR, Bohlman MC et al (2007) Water and salinity stress in grapevines: early and late changes in transcript and metabolite profiles. Funct Integr Genomics 7:111–134
Creelman RA, Mullet JE (1995) Jasmonic acid distribution and action in plants – regulation during development and response to biotic and abiotic stress. Proc Natl Acad Sci U S A 92:4114–4119

Dashnau JL (2007) Biological molecules as protectors against cellular damage caused by environmental extremes: the role of protectant–water interactions. Dissertations available from ProQuest. Paper AAI. http://repository.upenn.edu/dissertations/AAI3260893
De Micco V, Aronne G (2012) Morpho-anatomical traits for plant adaptation to drought. In: Aroca R (ed) Plant responses to drought stress. Springer, Berlin, pp 37–61
Degenkolbe T, Do PT, Zuther E, Repsilber D, Walther D, Hincha DK et al (2009) Expression profiling of rice cultivars differing in their tolerance to long-term drought stress. Plant Mol Biol 69:133–153
Den Herder G, Van Isterdael G, Beeckman T, De Smet I (2010) The roots of a new green revolution. Trends Plant Sci 15:600–607
Deokar AA, Kondawar V, Jain PK, Karuppayil SM, Raju NL, Vadez V et al (2011) Comparative analysis of expressed sequence tags (ESTs) between drought-tolerant and -susceptible genotypes of chickpea under terminal drought stress. BMC Plant Biol 11
Ding Y, Avramova Z, Fromm M (2011) The Arabidopsis trithorax-like factor ATX1 functions in dehydration stress responses via ABA-dependent and ABA-independent pathways. Plant J 66:735–744
Ding Y, Tao Y, Zhu C (2013) Emerging roles of microRNAs in the mediation of drought stress response in plants. J Exp Bot 64:3077–3086
Dixon HH, Joly J (1895) On the ascent of sap. Phil Trans R Soc Lond 186B:563–576
Djafi N, Vergnolle C, Cantrel C, Wietrzynski W, Delage E, Cochet F et al (2013) The Arabidopsis DREB2 genetic pathway is constitutively repressed by basal phosphoinositide-dependent phospholipase C coupled to diacylglycerol kinase. Front Plant Sci 4:307
Djilianov D, Ivanov S, Moyankova D, Miteva L, Kirova E, Alexieva V et al (2011) Sugar ratios, glutathione redox status and phenols in the resurrection species *Haberlea rhodopensis* and the closely related non-resurrection species *Chirita eberhardtii*. Plant Biol 13:767–776
Do PT, Degenkolbe T, Erban A, Heyer AG, Kopka J, Kohl KI et al (2013) Dissecting rice polyamine metabolism under controlled long-term drought stress. PLoS One 8, e60325
Ehleringer JR, Cooper TA (1992) On the role of orientation in reducing photoinhibitory damage in photosynthetic-twig desert shrubs. Plant Cell Environ 15:301–306
Ehleringer JR, Cooper TA (2006) On the role of orientation in reducing photoinhibitory damage in photosynthetic-twig desert shrubs. Plant Cell Environ Sci Technol 15:301–306
Eisenstein M (2013) Plant breeding discovery in a dry spell. Nature 501:S7–S9
Faria T, Silverio D, Breia E, Cabral R, Abadia A, Abadia J et al (1998) Differences in the response of carbon assimilation to summer stress (water deficits, high light and temperature) in four Mediterranean tree species. Physiol Plant 102:419–428
Feng HQ, Li HY, Sun K (2009) Enhanced expression of alternative oxidase genes is involved in the tolerance of rice (*Oryza sativa* L.) seedlings to drought stress. Zeitschr Naturforsch C J Biosci 64:704–710
Flexas J, Medrano H (2002) Drought-inhibition of photosynthesis in C3 plants: stomatal and non-stomatal limitations revisited. Ann Bot 89:183–189
Flexas J, Ribas-Carbo M, Bota J, Galmes J, Henkle M, Martinez-Canellas S et al (2006) Decreased Rubisco activity during water stress is not induced by decreased relative water content but related to conditions of low stomatal conductance and chloroplast CO2 concentration. New Phytol 172:73–82
Franks SJ (2011) Plasticity and evolution in drought avoidance and escape in the annual plant *Brassica rapa*. New Phytol 19:249–257
Franks PJ, Farquhar GD (2001) The effect of exogenous abscisic acid on stomatal development, stomatal mechanics, and leaf gas exchange in *Tradescantia virginiana*. Plant Physiol 125:935–942
Frost CJ, Nyamdari B, Tsai CJ, Harding SA (2012) The tonoplast-localized sucrose transporter in populus (PtaSUT4) regulates whole-plant water relations, responses to water stress, and photosynthesis. PLoS One 7, e44467
Fujii H, Zhu JK (2012) Osmotic stress signaling via protein kinases. Cell Mol Life Sci 69:3165–3173

Fujii H, Chinnusamy V, Rodrigues A, Rubio S, Antoni R, Park SY et al (2009) In vitro reconstitution of an abscisic acid signalling pathway. Nature 462:660–664

Fujita Y, Fujita M, Shinozaki K, Yamaguchi-Shinozaki K (2011) ABA-mediated transcriptional regulation in response to osmotic stress in plants. J Plant Res 124:509–525

Fukao T, Xiong LZ (2013) Genetic mechanisms conferring adaptation to submergence and drought in rice: simple or complex? Curr Opin Plant Biol 16:196–204

Galmes M (2007) Flexas. Photosynthesis and photoinhibition in response to drought in a pubescent (var. minor) and a glabrous (var. palaui) variety of *Digitalis minor*. Environ Exp Bot 60:105–1011

Galvez DA, Landhausser SM, Tyree MT (2013) Low root reserve accumulation during drought may lead to winter mortality in poplar seedlings. New Phytol 198:139–148

Garg AK, Kim JK, Owens TG, Ranwala AP, Do Choi Y, Kochian LV et al (2002) Trehalose accumulation in rice plants confers high tolerance levels to different abiotic stresses. Proc Natl Acad Sci U S A 99:15898–15903

Ge LF, Chao DY, Shi M, Zhu MZ, Gao JP, Lin HX (2008) Overexpression of the trehalose-6-phosphate phosphatase gene OsTPP1 confers stress tolerance in rice and results in the activation of stress responsive genes. Planta 228:191–201

Geber MA, Dawson TE (1997) Genetic variation in stomatal and biochemical limitations to photosynthesis in the annual plant, *Polygonum arenastrum*. Oecologia 109:535–546

Gechev TS, Dinakar C, Benina M, Toneva V, Bartels D (2012) Molecular mechanisms of desiccation tolerance in resurrection plants. Cell Mol Life Sci 69:3175–3186

Gechev TS, Benina M, Obata T, Tohge T, Sujeeth N, Minkov I et al (2013) Molecular mechanisms of desiccation tolerance in the resurrection glacial relic *Haberlea rhodopensis*. Cell Mol Life Sci 70:689–709

Geiger D, Scherzer S, Mumm P, Stange A, Marten I, Bauer H et al (2009) Activity of guard cell anion channel SLAC1 is controlled by drought-stress signaling kinase-phosphatase pair. Proc Natl Acad Sci U S A 106:21425–21430

Geiger D, Scherzer S, Mumm P, Marten I, Ache P, Matschi S et al (2010) Guard cell anion channel SLAC1 is regulated by CDPK protein kinases with distinct Ca2+ affinities. Proc Natl Acad Sci U S A 107:8023–8028

Gill SS, Tuteja N (2010) Polyamines and abiotic stress tolerance in plants. Plant Signal Behav 5:26–33

Giuliani S, Sanguineti MC, Tuberosa R, Bellotti M, Salvi S, Landi P (2005) Root-ABA1, a major constitutive QTL, affects maize root architecture and leaf ABA concentration at different water regimes. J Exp Bot 56:3061–3070

Gobert A, Isayenkov S, Voelker C, Czempinski K, Maathuis FJ (2007) The two-pore channel TPK1 gene encodes the vacuolar K+ conductance and plays a role in K+ homeostasis. Proc Natl Acad Sci U S A 104:10726–10731

Goldberg AD, Allis CD, Bernstein E (2007) Epigenetics: a landscape takes shape. Cell 128:635–638

Gonzalez RM, Ricardi MM, Iusem ND (2013) Epigenetic marks in an adaptive water stress-responsive gene in tomato roots under normal and drought conditions. Epigenetics 8:864–872

Granot G, Sikron-Persi N, Gaspan O, Florentin A, Talwara S, Paul LK et al (2009) Histone modifications associated with drought tolerance in the desert plant *Zygophyllum dumosum* Boiss. Planta 231:27–34

Gray JE, Holroyd GH, van der Lee FM, Bahrami AR, Sijmons PC, Woodward FI et al (2000) The HIC signalling pathway links CO2 perception to stomatal development. Nature 408:713–716

Groppa MD, Benavides MP (2008) Polyamines and abiotic stress: recent advances. Amino Acids 34:35–45

Guerra D, Mastrangelo AM, Lopez-Torrejon G, Marzin S, Schweizer P, Stanca AM et al (2012) Identification of a protein network interacting with TdRF1, a wheat RING ubiquitin ligase with a protective role against cellular dehydration. Plant Physiol 158:777–789

Hamanishi ET, Thomas BR, Campbell MM (2012) Drought induces alterations in the stomatal development program in *Populus*. J Exp Bot 63:4959–4971

Han SK, Wagner D (2013) Role of chromatin in water stress responses in plants. J Exp Bot
Hannah MA, Zuther E, Buchel K, Heyer AG (2006a) Transport and metabolism of raffinose family oligosaccharides in transgenic potato. J Exp Bot 57:3801–3811
Hannah MA, Wiese D, Freund S, Fiehn O, Heyer AG, Hincha DK (2006b) Natural genetic variation of freezing tolerance in arabidopsis. Plant Physiol 142:98–112
Harb A, Krishnan A, Ambavaram MMR, Pereira A (2010) Molecular and physiological analysis of drought stress in Arabidopsis reveals early responses leading to acclimation in plant growth. Plant Physiol 154:1254–1271
Hare PD, Cress WA, Van Staden J (1998) Dissecting the roles of osmolyte accumulation during stress. Plant Cell Environ 21:535–553
Hauser MT, Aufsatz W, Jonak C, Luschnig C (2011) Transgenerational epigenetic inheritance in plants. Biochim Biophys Acta 1809:459–468
Hayano-Kanashiro C, Calderon-Vazquez C, Ibarra-Laclette E, Herrera-Estrella L, Simpson J (2009) Analysis of gene expression and physiological responses in three Mexican maize landraces under drought stress and recovery irrigation. PLoS One 4
Hazen SP, Pathan MS, Sanchez A, Baxter I, Dunn M, Estes B et al (2005) Expression profiling of rice segregating for drought tolerance QTLs using a rice genome array. Funct Integr Genomics 5:104–116
Heschel MS, Riginos C (2005) Mechanisms of selection for drought stress tolerance and avoidance in *Impatiens capensis* (Balsaminaceae). Am J Bot 92:37–44
Hetherington AM (2001) Guard cell signaling. Cell 107:711–714
Himmelbach A, Yang Y, Grill E (2003) Relay and control of abscisic acid signaling. Curr Opin Plant Biol 6:470–479
Hincha DK (1989) Low concentrations of trehalose protect isolated thylakoids against mechanical freeze-thaw damage. Biochim Biophys Acta 987:231–234
Hirayama T, Shinozaki K (2010) Research on plant abiotic stress responses in the post-genome era: past, present and future. Plant J 61:1041–1052
Hoekstra FA (2005) Differential longevities in desiccated anhydrobiotic plant systems. Integr Comp Biol 45:725–733
Hoekstra FA, Golovina EA, Buitink J (2001) Mechanisms of plant desiccation tolerance. Trends Plant Sci 6:431–438
Holbrook NM, Zwieniccki MA (1999) Embolism repair and xylem tension: do we need a miracle? Plant Physiol 120:7–10
Hu HH, Dai MQ, Yao JL, Xiao BZ, Li XH, Zhang QF et al (2006) Overexpressing a NAM, ATAF, and CUC (NAC) transcription factor enhances drought resistance and salt tolerance in rice. Proc Natl Acad Sci U S A 103:12987–12992
Huang XY, Chao DY, Gao JP, Zhu MZ, Shi M, Lin HX (2009) A previously unknown zinc finger protein, DST, regulates drought and salt tolerance in rice via stomatal aperture control. Genes Dev 23:1805–1817
Jablonka E, Raz G (2009) Transgenerational epigenetic inheritance: prevalence, mechanisms, and implications for the study of heredity and evolution. Q Rev Biol 84:131–176
Jain NK, Roy I (2009) Effect of trehalose on protein structure. Protein Sci 18:24–36
Jang JY, Kim DG, Kim YO, Kim JS, Kang H (2004) An expression analysis of a gene family encoding plasma membrane aquaporins in response to abiotic stresses in *Arabidopsis thaliana*. Plant Mol Biol 54:713–725
Jarvis PG, Mcnaughton KG (1986) Stomatal control of transpiration – scaling up from leaf to region. Adv Ecol Res 15:1–49
Jensen WA, Salisbury FB (1972) Botany: an ecological approach. Wadsworth Publishing Company, Belmont, California
Jenuwein T, Allis CD (2001) Translating the histone code. Science 293:1074–1080
Jeong JS, Kim YS, Redillas MCFR, Jang G, Jung H, Bang SW et al (2013) OsNAC5 overexpression enlarges root diameter in rice plants leading to enhanced drought tolerance and increased grain yield in the field. Plant Biotechnol J 11:101–114

Ji KX, Wang YY, Sun WN, Lou QJ, Mei HW, Shen SH et al (2012) Drought-responsive mechanisms in rice genotypes with contrasting drought tolerance during reproductive stage. J Plant Physiol 169:336–344
Jin Y, Yang HX, Wei Z, Ma H, Ge XC (2013) Rice male development under drought stress: phenotypic changes and stage-dependent transcriptomic reprogramming. Mol Plant 6:1630–1645
Jonak C, Kiegerl S, Ligterink W, Barker PJ, Huskisson NS, Hirt H (1996) Stress signaling in plants: a mitogen-activated protein kinase pathway is activated by cold and drought. Proc Natl Acad Sci U S A 93:11274–11279
Juenger TE (2013) Natural variation and genetic constraints on drought tolerance. Curr Opin Plant Biol 16:274–281
Jung C, Seo JS, Han SW, Koo YJ, Kim CH, Song SI et al (2008) Overexpression of AtMYB44 enhances stomatal closure to confer abiotic stress tolerance in transgenic Arabidopsis. Plant Physiol 146:623–635
Kadioglu A, Terzi R, Saruhan N, Saglam A (2012) Current advances in the investigation of leaf rolling caused by biotic and abiotic stress factors. Plant Sci 182:42–48
Kakkar RK, Sawhney VK (2002) Polyamine research in plants – a changing perspective. Physiol Plant 116:281–292
Kaneda T, Taga Y, Takai R, Iwano M, Matsui H, Takayama S et al (2009) The transcription factor OsNAC4 is a key positive regulator of plant hypersensitive cell death. EMBO J 28:926–936
Kapazoglou A, Drosou V, Argiriou A, Tsaftaris AS (2013) The study of a barley epigenetic regulator, HvDME, in seed development and under drought. BMC Plant Biol 13:172
Karim S, Aronsson H, Ericson H, Pirhonen M, Leyman B, Welin B et al (2007) Improved drought tolerance without undesired side effects in transgenic plants producing trehalose. Plant Mol Biol 64:371–386
Kaufmann I, Schulze-Till T, Schneider HU, Zimmermann U, Jakob P, Wegner LH (2009) Functional repair of embolized vessels in maize roots after temporal drought stress, as demonstrated by magnetic resonance imaging. New Phytol 184:245–256
Khan M, Rozhon W, Bigeard J, Pflieger D, Husar S, Pitzschke A et al (2013) Brassinosteroid-regulated GSK3/Shaggy-like kinases phosphorylate mitogen-activated protein (MAP) kinase kinases, which control stomata development in *Arabidopsis thaliana*. J Biol Chem 288:7519–7527
Kim JM, To TK, Ishida J, Morosawa T, Kawashima M, Matsui A et al (2008) Alterations of lysine modifications on the histone H3 N-Tail under drought stress conditions in *Arabidopsis thaliana*. Plant Cell Physiol 49:1580–1588
Kim JM, To TK, Nishioka T, Seki M (2010a) Chromatin regulation functions in plant abiotic stress responses. Plant Cell Environ 33:604–611
Kim TH, Bohmer M, Hu HH, Nishimura N, Schroeder JI (2010b) Guard cell signal transduction network: advances in understanding abscisic acid, CO2, and Ca2+ signaling. Annu Rev Plant Biol 61:561–591
Kim JM, To TK, Ishida J, Matsui A, Kimura H, Seki M (2012a) Transition of chromatin status during the process of recovery from drought stress in *Arabidopsis thaliana*. Plant Cell Physiol 53:847–856
Kim T-W, Michniewicz M, Bergmann DC, Wang Z-Y (2012b) Brassinosteroid regulates stomatal development by GSK3-mediated inhibition of a MAPK pathway. Nature 482:419-U1526
Kim H, Lee K, Hwang H, Bhatnagar N, Kim DY, Yoon IS et al (2014) Overexpression of PYL5 in rice enhances drought tolerance, inhibits growth, and modulates gene expression. J Exp Bot 65:453–464
Kirch HH, Vera-Estrella R, Golldack D, Quigley F, Michalowski CB, Barkla BJ et al (2000) Expression of water channel proteins in *Mesembryanthemum crystallinum*. Plant Physiol 123:111–124
Kishor PBK, Hong ZL, Miao GH, Hu CAA, Verma DPS (1995) Overexpression of delta-pyrroline-5-carboxylate synthetase increases proline production and confers osmotolerance in transgenic plants. Plant Physiol 108:1387–1394

Kishor PBK, Sangam S, Amrutha RN, Laxmi PS, Naidu KR, Rao KRSS et al (2005) Regulation of proline biosynthesis, degradation, uptake and transport in higher plants: its implications in plant growth and abiotic stress tolerance. Curr Sci 88:424–438
Koster KL, Leopold AC (1988) Sugars and desiccation tolerance in seeds. Plant Physiol 88:829–832
Koster KL, Balsamo RA, Espinoza C, Oliver MJ (2010) Desiccation sensitivity and tolerance in the moss *Physcomitrella patens*: assessing limits and damage. Plant Growth Regul 62:293–302
Krasensky J, Jonak C (2012) Drought, salt, and temperature stress-induced metabolic rearrangements and regulatory networks. J Exp Bot 63:1593–1608
Kushiro T, Okamoto M, Nakabayashi K, Yamagishi K, Kitamura S, Asami T et al (2004) The Arabidopsis cytochrome P450CYP707A encodes ABA 8′-hydroxylases: key enzymes in ABA catabolism. Embo J 23:1647–1656
Laird PW (2010) Principles and challenges of genome-wide DNA methylation analysis. Nat Rev Genet 11:191–203
Larcher W (2000) Temperature stress and survival ability of Mediterranean sclerophyllous plants. Plant Biosyst 134:279–295
Lata C, Prasad M (2011) Role of DREBs in regulation of abiotic stress responses in plants. J Exp Bot 62:4731–4748
Le TN, McQueen-Mason SJ (2007) Desiccation-tolerant plants in dry environments. In: Amils R, Ellis-Evans C, Hinghofer-Szalkay H (eds) Life in extreme environments. Springer, Amsterdam, pp 265–275
Lee SC, Lan W, Buchanan BB, Luan S (2009) A protein kinase-phosphatase pair interacts with an ion channel to regulate ABA signaling in plant guard cells. Proc Natl Acad Sci U S A 106:21419–21424
Lee SJ, Kang JY, Park HJ, Kim MD, Bae MS, Choi HI et al (2010) DREB2C interacts with ABF2, a bZIP protein regulating abscisic acid-responsive gene expression, and its overexpression affects abscisic acid sensitivity. Plant Physiol 153:716–727
Legnaioli T, Cuevas J, Mas P (2009) TOC1 functions as a molecular switch connecting the circadian clock with plant responses to drought. Embo J 28:3745–3757
Lelievre F, Seddaiu G, Ledda L, Porqueddu C, Volaire F (2011) Water use efficiency and drought survival in Mediterranean perennial forage grasses. Field Crop Res 121:333–342
Lens F, Tixier A, Cochard H, Sperry JS, Jansen S, Herbette S (2013) Embolism resistance as a key mechanism to understand adaptive plant strategies. Curr Opin Plant Biol 16:287–292
Levitt J (1980) Responses of plants to environmental stresses. Academic, New York
Li XY, Wang XF, He K, Ma YQ, Su N, He H et al (2008) High-resolution mapping of epigenetic modifications of the rice genome uncovers interplay between DNA methylation, histone methylation, and gene expression. Plant Cell 20:259–276
Li HW, Zang BS, Deng XW, Wang XP (2011) Overexpression of the trehalose-6-phosphate synthase gene OsTPS1 enhances abiotic stress tolerance in rice. Planta 234:1007–1018
Liang YK, Dubos C, Dodd IC, Holroyd GH, Hetherington AM, Campbell MM (2005) AtMYB61, an R2R3-MYB transcription factor controlling stomatal aperture in *Arabidopsis thaliana*. Curr Biol 15:1201–1206
Liu SW, Jiang YW (2010) Identification of differentially expressed genes under drought stress in perennial ryegrass. Physiol Plant 139:375–387
Liu F, Stützel H (2004) Biomass partitioning, specific leaf area, and water use efficiency of vegetable amaranth (*Amaranthus* spp.) in response to drought stress. Sci Hortic 102:15–27
Liu Q, Kasuga M, Sakuma Y, Abe H, Miura S, Yamaguchi-Shinozaki K et al (1998) Two transcription factors, DREB1 and DREB2, with an EREBP/AP2 DNA binding domain separate two cellular signal transduction pathways in drought- and low-temperature-responsive gene expression, respectively, in Arabidopsis. Plant Cell 10:1391–1406
Liu H-H, Tian X, Li Y-J, Wu C-A, Zheng C-C (2008) Microarray-based analysis of stress-regulated microRNAs in Arabidopsis thaliana. RNA (New York, NY) 14:836–843

Liu GZ, Li XL, Jin SX, Liu XY, Zhu LF, Nie YC et al (2014) Overexpression of rice NAC Gene SNAC1 improves drought and salt tolerance by enhancing root development and reducing transpiration rate in transgenic cotton. PLoS One 9

Lorenz WW, Alba R, Yu YS, Bordeaux JM, Simoes M, Dean JFD (2011) Microarray analysis and scale-free gene networks identify candidate regulators in drought-stressed roots of loblolly pine (*P. taeda* L.). BMC Genomics 12

Lorenzo O, Solano R (2005) Molecular players regulating the jasmonate signalling network. Curr Opin Plant Biol 8:532–540

Ludlow MM (1989) Strategies of response to water stress. In: Kreeb KH, Richter H, Hinckley TM (eds) Structural and functional responses to water stress. SPB Academic Press, The Hague

Ludlow MM, Ng TT (1974) Water stress suspends leaf aging. Plant Sci Lett 3:235–240

Lyzenga WJ, Stone SL (2012) Abiotic stress tolerance mediated by protein ubiquitination. J Exp Bot 63:599–616

Ma Y, Szostkiewicz I, Korte A, Moes D, Yang Y, Christmann A et al (2009) Regulators of PP2C phosphatase activity function as abscisic acid sensors. Science 324:1064–1068

Madlung A, Comai L (2004) The effect of stress on genome regulation and structure. Ann Bot 94:481–495

Mahdieh M, Mostajeran A, Horie T, Katsuhara M (2008) Drought stress alters water relations and expression of PIP-type aquaporin genes in *Nicotiana tabacum* plants. Plant Cell Physiol 49:801–813

Maroco JP, Pereira JS, Chaves MM (2000) Growth, photosynthesis and water-use efficiency of two C-4 Sahelian grasses subjected to water deficits. J Arid Environ 45:119–137

Matsukura S, Mizoi J, Yoshida T, Todaka D, Ito Y, Maruyama K et al (2010) Comprehensive analysis of rice DREB2-type genes that encode transcription factors involved in the expression of abiotic stress-responsive genes. Mol Genet Genomics 283:185–196

Mattioli R, Costantino P, Trovato M (2009) Proline accumulation in plants: not only stress. Plant Signal Behav 4:1016–1018

Maurel C, Verdoucq L, Luu DT, Santoni V (2008) Plant aquaporins: membrane channels with multiple integrated functions. Annu Rev Plant Biol 59:595–624

Mcclintock B (1984) The significance of responses of the genome to challenge. Science 226:792–801

McDonald AJS, Davies WJ (1996) Keeping in touch: responses of the whole plant to deficits in water and nitrogen supply. Adv Bot Res 22:229–300

McDowell NG (2011) Mechanisms linking drought, hydraulics, carbon metabolism, and vegetation mortality. Plant Physiol 155:1051–1059

McDowell N, Pockman WT, Allen CD, Breshears DD, Cobb N, Kolb T et al (2008) Mechanisms of plant survival and mortality during drought: why do some plants survive while others succumb to drought? New Phytol 178:719–739

McKay JKJ, Richards JHJ, Mitchell-Olds TT (2003) Genetics of drought adaptation in *Arabidopsis thaliana*: I. Pleiotropy contributes to genetic correlations among ecological traits. Mol Ecol 12:1137–1151

Mir RR, Zaman-Allah M, Sreenivasulu N, Trethowan R, Varshney RK (2012) Integrated genomics, physiology and breeding approaches for improving drought tolerance in crops. Theor Appl Genet 125:625–645

Mittler R (2006) Abiotic stress, the field environment and stress combination. Trends Plant Sci 11:15–19

Mizoi J, Shinozaki K, Yamaguchi-Shinozaki K (2012) AP2/ERF family transcription factors in plant abiotic stress responses. Biochim Biophys Acta 1819:86–96

Mooney HA, Pearcy RW, Ehleringer J (1987) Plant physiological ecology today. Bioscience 37:18–20

Moore JP, Westall KL, Ravenscroft N, Farrant JM, Lindsey GG, Brandt WF (2005) The predominant polyphenol in the leaves of the resurrection plant *Myrothamnus flabellifolius*, 3,4,5 tri-O-galloylquinic acid, protects membranes against desiccation and free radical-induced oxidation. Biochem J 385:301–308

Munemasa S, Muroyama D, Nagahashi H, Nakamura Y, Mori IC, Murata Y (2013) Regulation of reactive oxygen species-mediated abscisic acid signaling in guard cells and drought tolerance by glutathione. Front Plant Sci 4:472

Munns R, Sharp RE (1993) Involvement of abscisic-acid in controlling plant-growth in soils of low water potential. Aust J Plant Physiol 20:425–437

Murata N, Takahashi S, Nishiyama Y, Allakhverdiev SI (1767) Photoinhibition of photosystem II under environmental stress. Biochim Biophys Acta 2007:414–421

Mustilli AC, Merlot S, Vavasseur A, Fenzi F, Giraudat J (2002) Arabidopsis OST1 protein kinase mediates the regulation of stomatal aperture by abscisic acid and acts upstream of reactive oxygen species production. Plant Cell 14:3089–3099

Nagy R, Grob H, Weder B, Green P, Klein M, Frelet-Barrand A et al (2009) The Arabidopsis ATP-binding cassette protein AtMRP5/AtABCC5 is a high affinity inositol hexakisphosphate transporter involved in guard cell signaling and phytate storage. J Biol Chem 284:33614–33622

Nakashima K, Shinwari ZK, Sakuma Y, Seki M, Miura S, Shinozaki K et al (2000) Organization and expression of two Arabidopsis DREB2 genes encoding DRE-binding proteins involved in dehydration- and high-salinity-responsive gene expression. Plant Mol Biol 42:657–665

Nakashima K, Ito Y, Yamaguchi-Shinozaki K (2009) Transcriptional regulatory networks in response to abiotic stresses in Arabidopsis and grasses. Plant Physiol 149:88–95

Nardini A, Lo Gullo MA, Salleo S (2011) Refilling embolized xylem conduits: is it a matter of phloem unloading? Plant Sci 180:604–611

Negi J, Moriwaki K, Konishi M, Yokoyama R, Nakano T, Kusumi K et al (2013) A Dof transcription factor, SCAP1, is essential for the development of functional stomata in Arabidopsis. Curr Biol 23:479–484

Niederhuth CE, Schmitz RJ (2014) Covering your bases: inheritance of DNA methylation in plant genomes. Mol Plant 7:472–480

Nilson SE, Assmann SM (2010) The alpha-subunit of the Arabidopsis heterotrimeric G protein, GPA1, is a regulator of transpiration efficiency. Plant Physiol 152:2067–2077

Ning J, Li X, Hicks LM, Xiong L (2010) A Raf-like MAPKKK gene DSM1 mediates drought resistance through reactive oxygen species scavenging in rice. Plant Physiol 152:876–890

North HM, De Almeida A, Boutin JP, Frey A, To A, Botran L et al (2007) The Arabidopsis ABA-deficient mutant aba4 demonstrates that the major route for stress-induced ABA accumulation is via neoxanthin isomers. Plant J 50:810–824

Oertli JJ, Lips SH, Agami M (1990) The strength of sclerophyllous cells to resist collapse due to negative Turgor pressure. Acta Oecol Int J Ecol 11:281–289

Oliver MJ (2008) Biochemical and molecular mechanisms of desiccation tolerance in bryophytes. In: Goffinet B, Shaw AJ (eds) Bryophyte biology, 2nd edn. Cambridge University Press, New York, NY, pp 269–298

Oliver MJ, Jain R, Balbuena TS, Agrawal G, Gasulla F, Thelen JJ (2011) Proteome analysis of leaves of the desiccation-tolerant grass, *Sporobolus stapfianus*, in response to dehydration. Phytochemistry 72:1273–1284

Oono Y, Seki M, Nanjo T, Narusaka M, Fujita M, Satoh R et al (2003) Monitoring expression profiles of Arabidopsis gene expression during rehydration process after dehydration using ca. 7000 full-length cDNA microarray. Plant J 34:868–887

Otto T, Zoitan T, Scott P (2009) Vegetative desiccation tolerance: is it a goldmine for bioengineering crops? Plant Sci 176:187–199

Ouyang SQ, Liu YF, Liu P, Lei G, He SJ, Ma B et al (2010) Receptor-like kinase OsSIK1 improves drought and salt stress tolerance in rice (*Oryza sativa*) plants. Plant J 62:316–329

Ozturk ZN, Talame V, Deyholos M, Michalowski CB, Galbraith DW, Gozukirmizi N et al (2002) Monitoring large-scale changes in transcript abundance in drought- and salt-stressed barley. Plant Mol Biol 48:551–573

Pandey GK, Grant JJ, Cheong YH, Kim BG, Li LG, Luan S (2005) ABR1, an APETALA2-domain transcription factor that functions as a repressor of ABA response in Arabidopsis. Plant Physiol 139:1185–1193

Papaefthimiou D, Tsaftaris A (2012) Characterization of a drought inducible trithorax-like H3K4 methyltransferase from barley. Biol Plant 56:683–692

Park SY, Fung P, Nishimura N, Jensen DR, Fujii H, Zhao Y et al (2009) Abscisic acid inhibits type 2C protein phosphatases via the PYR/PYL family of START proteins. Science 324:1068–1071

Pastore D, Trono D, Laus MN, Di Fonzo N, Flagella Z (2007) Possible plant mitochondria involvement in cell adaptation to drought stress – a case study: durum wheat mitochondria. J Exp Bot 58:195–210

Paul MJ, Primavesi LF, Jhurreea D, Zhang YH (2008) Trehalose metabolism and signaling. Annu Rev Plant Biol 59:417–441

Pecinka A, Scheid OM (2012) Stress-induced chromatin changes: a critical view on their heritability. Plant Cell Physiol 53:801–808

Pei ZM, Murata Y, Benning G, Thomine S, Klusener B, Allen GJ et al (2000) Calcium channels activated by hydrogen peroxide mediate abscisic acid signalling in guard cells. Nature 406:731–734

Peng H, Zhang J (2009) Plant genomic DNA methylation in response to stresses: potential applications and challenges in plant breeding. Prog Nat Sci 19:1037–1045

Perera IY, Hung CY, Moore CD, Stevenson-Paulik J, Boss WF (2008) Transgenic Arabidopsis plants expressing the type 1 inositol 5-phosphatase exhibit increased drought tolerance and altered abscisic acid signaling. Plant Cell 20:2876–2893

Peters S, Mundree SG, Thomson JA, Farrant JM, Keller F (2007) Protection mechanisms in the resurrection plant *Xerophyta viscosa* (Baker): both sucrose and raffinose family oligosaccharides (RFOs) accumulate in leaves in response to water deficit. J Exp Bot 58:1947–1956

Pillitteri LJ, Torii KU (2012) Mechanisms of stomatal development. Annu Rev Plant Biol 63:591–614

Poorter H, Niklas KJ, Reich PB, Oleksyn J, Poot P, Mommer L (2012) Biomass allocation to leaves, stems and roots: meta-analyses of interspecific variation and environmental control. New Phytol 193:30–50

Porembski S (2011) Evolution, diversity, and habitats of Poikilohydrous vascular plants. *Plant desiccation tolerance*. Springer, Berlin, pp 139–156

Qin F, Sakuma Y, Tran LSP, Maruyama K, Kidokoro S, Fujita Y et al (2008) Arabidopsis DREB2A-interacting proteins function as RING E3 ligases and negatively regulate plant drought stress-responsive gene expression. Plant Cell 20:1693–1707

Rabbani MA, Maruyama K, Abe H, Khan MA, Katsura K, Ito Y et al (2003) Monitoring expression profiles of rice genes under cold, drought, and high-salinity stresses and abscisic acid application using cDNA microarray and RNA get-blot analyses. Plant Physiol 133:1755–1767

Raghavendra AS, Reddy KB (1987) Action of proline on stomata differs from that of abscisic-acid, G-substances, or methyl jasmonate. Plant Physiol 83:732–734

Riboni M, Galbiati M, Tonelli C, Conti L (2013) GIGANTEA enables drought escape response via abscisic acid-dependent activation of the florigens and suppressor of overexpression of CONSTANS1. Plant Physiol 162:1706–1719

Rodrigues ML, Pacheco CMA, Chaves MM (1995) Soil-plant water relations, root distribution and biomass partitioning in *Lupinus albus* L. under drought conditions. J Exp Bot 46:947–956

Rodrigues A, Adamo M, Crozet P, Margalha L, Confraria A, Martinho C et al (2013) ABI1 and PP2CA phosphatases are negative regulators of Snf1-related protein kinase1 signaling in Arabidopsis. Plant Cell 25:3871–3884

Rolland F, Baena-Gonzalez E, Sheen J (2006) Sugar sensing and signaling in plants: conserved and novel mechanisms. Annu Rev Plant Biol 57:675–709

Sagi M, Fluhr R (2006) Production of reactive oxygen species by plant NADPH oxidases. Plant Physiol 141:336–340

Saibo NJM, Lourenco T, Oliveira MM (2009) Transcription factors and regulation of photosynthetic and related metabolism under environmental stresses. Ann Bot 103:609–623

Saijo Y, Hata S, Kyozuka J, Shimamoto K, Izui K (2000) Over-expression of a single Ca2+-dependent protein kinase confers both cold and salt/drought tolerance on rice plants. Plant J 23:319–327
Sakuma Y, Maruyama K, Qin F, Osakabe Y, Shinozaki K, Yamaguchi-Shinozaki K (2006a) Dual function of an Arabidopsis transcription factor DREB2A in water-stress-responsive and heat-stress-responsive gene expression. Proc Natl Acad Sci U S A 103:18822–18827
Sakuma Y, Maruyama K, Osakabe Y, Qin F, Seki M, Shinozaki K et al (2006b) Functional analysis of an Arabidopsis transcription factor, DREB2A, involved in drought-responsive gene expression. Plant Cell 18:1292–1309
Salleo S, Trifilo P, Lo Cullo MA (2008) Vessel wall vibrations: trigger for embolism repair? Funct Plant Biol 35:289–297
Sanchez A, Shin J, Davis SJ (2011) Abiotic stress and the plant circadian clock. Plant Signal Behav 6:223–231
Santos AP, Serra T, Figueiredo DD, Barros P, Lourenço T, Chander S et al (2011) Transcription regulation of abiotic stress responses in rice: a combined action of transcription factors and epigenetic mechanisms. Omics J Integr Biol 15:839–857
Sapeta H, Costa JM, Lourenço T, Maroco J, van der Linde P, Oliveira MM (2013) Drought stress response in *Jatropha curcas*: growth and physiology. Environ Exp Bot 85:76–84
Sapeta H, Lourenço T, Lorenz S, Grumaz C, Kirstahler P, Barros PM, Costa JM, Sohn K, Oliveira MM (2015) Transcriptomics and physiological analyses reveal co-ordinated alteration of metabolic pathways in *Jatropha curcas* drought tolerance. J Exp Bot. DOI: 10.1093/jxb/erv499 First published online: November 23, 2015
Sato A, Sato Y, Fukao Y, Fujiwara M, Umezawa T, Shinozaki K et al (2009) Threonine at position 306 of the KAT1 potassium channel is essential for channel activity and is a target site for ABA-activated SnRK2/OST1/SnRK2.6 protein kinase. Biochem J 424:439–448
Saze H (2008) Epigenetic memory transmission through mitosis and meiosis in plants. Semin Cell Dev Biol 19:527–536
Saze H (2012) Transgenerational inheritance of induced changes in the epigenetic state of chromatin in plants. Genes Genet Syst 87:145–152
Schulze ED, Robichaux RH, Grace J, Rundel PW, Ehleringer JR (1987) Plant water balance. BioScience 37:30–37
Schurr U, Heckenberger U, Herdel K, Walter A, Feil R (2000) Leaf development in *Ricinus communis* during drought stress: dynamics of growth processes, of cellular structure and of sink-source transition. J Exp Bot 51:1515–1529
Secchi F, Zwieniecki MA (2011) Sensing embolism in xylem vessels: the role of sucrose as a trigger for refilling. Plant Cell Environ Sci Technol 34:514–524
Secchi F, Gilbert ME, Zwieniecki MA (2011) Transcriptome response to embolism formation in stems of *Populus trichocarpa* provides insight into signaling and the biology of refilling. Plant Physiol 157:1419–1429
Seki M, Narusaka M, Ishida J, Nanjo T, Fujita M, Oono Y et al (2002) Monitoring the expression profiles of 7000 Arabidopsis genes under drought, cold and high-salinity stresses using a full-length cDNA microarray. Plant J 31:279–292
Selvaraj MG, Ogawa S, Ishitani M (2013) Root phenomics-new windows to understand plant performance and increase crop productivity. J Plant Biochem Physiol 1:116
Seo M, Koshiba T (2011) Transport of ABA from the site of biosynthesis to the site of action. J Plant Res 124:501–507
Sherrard ME, Maherali H (2006) The adaptive significance of drought escape in *Avena barbata*, an annual grass. Evolution 60:2478–2489
Shinozaki K, Yamaguchi-Shinozaki K (1997) Gene expression and signal transduction in water-stress response. Plant Physiol 115:327–334
Shinozaki K, Yamaguchi-Shinozaki K (2007) Gene networks involved in drought stress response and tolerance. J Exp Bot 58:221–227

Shuai P, Liang D, Zhang Z, Yin W, Xia X (2013) Identification of drought-responsive and novel *Populus trichocarpa* microRNAs by high-throughput sequencing and their targets using degradome analysis. BMC Genomics 14:233
Sinha AK, Jaggi M, Raghuram B, Tuteja N (2011) Mitogen-activated protein kinase signaling in plants under abiotic stress. Plant Signal Behav 6:196–203
Sirichandra C, Gu D, Hu HC, Davanture M, Lee S, Djaoui M et al (2009) Phosphorylation of the Arabidopsis AtrbohF NADPH oxidase by OST1 protein kinase. FEBS Lett 583:2982–2986
Skirycz A, Claeys H, De Bodt S, Oikawa A, Shinoda S, Andriankaja M et al (2011) Pause-and-stop: the effects of osmotic stress on cell proliferation during early leaf development in Arabidopsis and a role for ethylene signaling in cell cycle arrest. Plant Cell 23:1876–1888
Slaughter A, Daniel X, Flors V, Luna E, Hohn B, Mauch-Mani B (2012) Descendants of primed Arabidopsis plants exhibit resistance to biotic stress. Plant Physiol 158:835–843
Sperry JS (2003) Evolution of water transport and xylem structure. Int J Plant Sci 164:S115–S127
Sridha S, Wu KQ (2006) Identification of AtHD2C as a novel regulator of abscisic acid responses in Arabidopsis. Plant J 46:124–133
Stark LR, Greenwood JL, Brinda JC, Oliver MJ (2013) The desert moss *Pterygoneurum lamellatum* (Pottiaceae) exhibits an inducible ecological strategy of desiccation tolerance: effects of rate of drying on shoot damage and regeneration. Am J Bot 100:1522–1531
Steudle E (2001) The cohesion-tension mechanism and the acquisition of water by plant roots. Annu Rev Plant Physiol Plant Mol Biol 52:847–875
Stone SL, Williams LA, Farmer LM, Vierstra RD, Callis J (2006) KEEP ON GOING, a RING E3 ligase essential for Arabidopsis growth and development, is involved in abscisic acid signaling. Plant Cell 18:3415–3428
Stoop JMH, Williamson JD, Pharr DM (1996) Mannitol metabolism in plants: a method for coping with stress. Trends Plant Sci 1:139–144
Strahl BD, Allis CD (2000) The language of covalent histone modifications. Nature 403:41–45
Suhita D, Raghavendra AS, Kwak JM, Vavasseur A (2004) Cytoplasmic alkalization precedes reactive oxygen species production during methyl jasmonate- and abscisic acid-induced stomatal closure. Plant Physiol 134:1536–1545
Suzuki N, Miller G, Morales J, Shulaev V, Torres MA, Mittler R (2011) Respiratory burst oxidases: the engines of ROS signaling. Curr Opin Plant Biol 14:691–699
Szabados L, Savoure A (2010) Proline: a multifunctional amino acid. Trends Plant Sci 15:89–97
Taji T, Ohsumi C, Iuchi S, Seki M, Kasuga M, Kobayashi M et al (2002) Important roles of drought- and cold-inducible genes for galactinol synthase in stress tolerance in *Arabidopsis thaliana*. Plant J 29:417–426
Talame V, Ozturk NZ, Bohnert HJ, Tuberosa R (2007) Barley transcript profiles under dehydration shock and drought stress treatments: a comparative analysis. J Exp Bot 58:229–240
Tanaka Y, Nose T, Jikumaru Y, Kamiya Y (2013) ABA inhibits entry into stomatal-lineage development in Arabidopsis leaves. Plant J 74:448–457
Tang S, Liang HY, Yan DH, Zhao Y, Han X, Carlson JE et al (2013) *Populus euphratica*: the transcriptomic response to drought stress. Plant Mol Biol 83:539–557
Tricker PJ, Gibbings JG, Lopez CMR, Hadley P, Wilkinson MJ (2012) Low relative humidity triggers RNA-directed de novo DNA methylation and suppression of genes controlling stomatal development. J Exp Bot 63:3799–3813
Tricker P, Rodriguez Lopez C, Hadley P, Wagstaff C, Wilkinson M (2013) Pre-conditioning the epigenetic response to high vapor pressure deficit increases the drought tolerance of *Arabidopsis thaliana*. Plant Signal Behav 8, e25974
Trindade I, Capitao C, Dalmay T, Fevereiro MP, dos Santos DM (2010) miR398 and miR408 are up-regulated in response to water deficit in *Medicago truncatula*. Planta 231:705–716
Tyerman SD, Niemietz CM, Bramley H (2002) Plant aquaporins: multifunctional water and solute channels with expanding roles. Plant Cell Environ 25:173–194
Tyree MT (1997) The cohesion-tension theory of sap ascent: current controversies. J Exp Bot 48:1753–1765

Ueda A, Kathiresan A, Inada M, Narita Y, Nakamura T, Shi WM et al (2004) Osmotic stress in barley regulates expression of a different set of genes than salt stress does. J Exp Bot 55:2213–2218

Uga Y, Sugimoto K, Ogawa S, Rane J, Ishitani M, Hara N et al (2013) Control of root system architecture by DEEPER ROOTING 1 increases rice yield under drought conditions. Nat Genet 45:1097

Umezawa T, Sugiyama N, Mizoguchi M, Hayashi S, Myouga F, Yamaguchi-Shinozaki K et al (2009) Type 2C protein phosphatases directly regulate abscisic acid-activated protein kinases in Arabidopsis. Proc Natl Acad Sci U S A 106:17588–17593

Urao T, Yakubov B, Satoh R, Yamaguchi-Shinozaki K, Seki M, Hirayama T et al (1999) A transmembrane hybrid-type histidine kinase in Arabidopsis functions as an osmosensor. Plant Cell 11:1743–1754

Valerio C, Costa A, Marri L, Issakidis-Bourguet E, Pupillo P, Trost P et al (2011) Thioredoxin-regulated beta-amylase (BAM1) triggers diurnal starch degradation in guard cells, and in mesophyll cells under osmotic stress. J Exp Bot 62:545–555

Valliyodan B, Nguyen HT (2006) Understanding regulatory networks and engineering for enhanced drought tolerance in plants. Curr Opin Plant Biol 9:189–195

van Dijk K, Ding Y, Malkaram S, Riethoven JJM, Liu R, Yang JY et al (2010) Dynamic changes in genome-wide histone H3 lysine 4 methylation patterns in response to dehydration stress in *Arabidopsis thaliana*. BMC Plant Biol 10:238

Vanlerberghe GC (2013) Alternative oxidase: a mitochondrial respiratory pathway to maintain metabolic and signaling homeostasis during abiotic and biotic stress in plants. Int J Mol Sci 14:6805–6847

Verbruggen N, Hermans C (2008) Proline accumulation in plants: a review. Amino Acids 35:753–759

Vlad F, Rubio S, Rodrigues A, Sirichandra C, Belin C, Robert N et al (2009) Protein phosphatases 2C regulate the activation of the Snf1-related kinase OST1 by abscisic acid in Arabidopsis. Plant Cell 21:3170–3184

Volaire F, Norton M (2006) Summer dormancy in perennial temperate grasses. Ann Bot 98:927–933

Volaire F, Seddaiu G, Ledda L, Lelievre F (2009) Water deficit and induction of summer dormancy in perennial Mediterranean grasses. Ann Bot 103:1337–1346

Wang J, Vanlerberghe GC (2013) A lack of mitochondrial alternative oxidase compromises capacity to recover from severe drought stress. Physiol Plant 149:461–473

Wang XQ, Ullah H, Jones AM, Assmann SM (2001) G protein regulation of ion channels and abscisic acid signaling in Arabidopsis guard cells. Science 292:2070–2072

Wang WS, Pan YJ, Zhao XQ, Dwivedi D, Zhu LH, Ali J et al (2011a) Drought-induced site-specific DNA methylation and its association with drought tolerance in rice (*Oryza sativa* L.). J Exp Bot 62:1951–1960

Wang D, Pan YJ, Zhao XQ, Zhu LH, Fu BY, Li ZK (2011b) Genome-wide temporal-spatial gene expression profiling of drought responsiveness in rice. BMC Genomics 12

Wasilewska A, Vlad F, Sirichandra C, Redko Y, Jammes F, Valon C et al (2008) An update on abscisic acid signaling in plants and more. Mol Plant 1:198–217

Watt M, Moosavi S, Cunningham SC, Kirkegaard JA, Rebetzke GJ, Richards RA (2013) A rapid, controlled-environment seedling root screen for wheat correlates well with rooting depths at vegetative, but not reproductive, stages at two field sites. Ann Bot 112:447–455

Weng X, Wang L, Wang J, Hu Y, Du H, Xu C et al (2014) Grain number, plant height, and heading date7 is a central regulator of growth, development, and stress response. Plant Physiol 164:735–747

Witzany G (2006) Plant communication from biosemiotic perspective: differences in abiotic and biotic signal perception determine content arrangement of response behavior. Context determines meaning of meta-, inter- and intraorganismic plant signaling. Plant Signal Behav 1:169–178

Wu YP, Hu XW, Wang YR (2009) Growth, water relations, and stomatal development of *Caragana korshinskii* Kom. and *Zygophyllum xanthoxylum* (Bunge) Maxim. seedlings in response to water deficits. New Zeal J Agric Res 52:185–193
Wu CA, Lowry DB, Nutter LI, Willis JH (2010) Natural variation for drought-response traits in the *Mimulus guttatus* species complex. Oecologia 162:23–33
Xiang Y, Huang Y, Xiong L (2007) Characterization of stress-responsive CIPK genes in rice for stress tolerance improvement. Plant Physiol 144:1416–1428
Xie GS, Kato H, Sasaki K, Imai R (2009) A cold-induced thioredoxin h of rice, OsTrx23, negatively regulates kinase activities of OsMPK3 and OsMPK6 in vitro. FEBS Lett 583:2734–2738
Xiong L, Yang Y (2003) Disease resistance and abiotic stress tolerance in rice are inversely modulated by an abscisic acid-inducible mitogen-activated protein kinase. Plant Cell 15:745–759
Xiong LM, Wang RG, Mao GH, Koczan JM (2006) Identification of drought tolerance determinants by genetic analysis of root response to drought stress and abscisic acid. Plant Physiol 142:1065–1074
Xu YJ, Gao S, Yang YJ, Huang MY, Cheng LN, Wei Q et al (2013) Transcriptome sequencing and whole genome expression profiling of chrysanthemum under dehydration stress. BMC Genomics 14
Xu MY, Zhang L, Li WW, Hu XL, Wang MB, Fan YL et al (2014) Stress-induced early flowering is mediated by miR169 in *Arabidopsis thaliana*. J Exp Bot 65:89–101
Xue GP, McIntyre CL, Glassop D, Shorter R (2008) Use of expression analysis to dissect alterations in carbohydrate metabolism in wheat leaves during drought stress. Plant Mol Biol 67:197–214
Yamada M, Morishita H, Urano K, Shiozaki N, Yamaguchi-Shinozaki K, Shinozaki K et al (2005) Effects of free proline accumulation in petunias under drought stress. J Exp Bot 56:1975–1981
Yamaguchi-Shinozaki K, Shinozaki K (1994) A novel cis-acting element in an Arabidopsis gene is involved in responsiveness to drought, low-temperature, or high-salt stress. Plant Cell 6:251–264
Yan DH, Fenning T, Tang S, Xia XL, Yin WL (2012) Genome-wide transcriptional response of *Populus euphratica* to long-term drought stress. Plant Sci 195:24–35
Yancey PH (2005) Organic osmolytes as compatible, metabolic and counteracting cytoprotectants in high osmolarity and other stresses. J Exp Biol 208:2819–2830
Yang J, Zhang J, Huang Z, Zhu Q, Wang L (2000) Remobilization of carbon reserves is improved by controlled soil-drying during grain filling of wheat. Crop Sci 40:1645–1655
Yang J, Zhang J, Wang Z, Zhu Q, Wang W (2001) Remobilization of carbon reserves in response to water deficit during grain filling of rice. Field Crop Res 71:47–55
Yang JC, Zhang JH, Liu K, Wang ZQ, Liu LJ (2007) Involvement of polyamines in the drought resistance of rice. J Exp Bot 58:1545–1555
Yeo ET, Kwon HB, Han SE, Lee JT, Ryu JC, Byun MO (2000) Genetic engineering of drought resistant potato plants by introduction of the trehalose-6-phosphate synthase (TPS1) gene from *Saccharomyces cerevisiae*. Mol Cells 10:263–268
Yoshiba Y, Kiyosue T, Nakashima K, YamaguchiShinozaki K, Shinozaki K (1997) Regulation of levels of proline as an osmolyte in plants under water stress. Plant Cell Physiol 38:1095–1102
Yoshida R, Umezawa T, Mizoguchi T, Takahashi S, Takahashi F, Shinozaki K (2006) The regulatory domain of SRK2E/OST1/SnRK2.6 interacts with ABI1 and integrates abscisic acid (ABA) and osmotic stress signals controlling stomatal closure in Arabidopsis. J Biol Chem 281:5310–5318
Yoshida T, Fujita Y, Sayama H, Kidokoro S, Maruyama K, Mizoi J et al (2010) AREB1, AREB2, and ABF3 are master transcription factors that cooperatively regulate ABRE-dependent ABA signaling involved in drought stress tolerance and require ABA for full activation. Plant J 61:672–685

Zeller G, Henz SR, Widmer CK, Sachsenberg T, Ratsch G, Weigel D et al (2009) Stress-induced changes in the *Arabidopsis thaliana* transcriptome analyzed using whole-genome tiling arrays. Plant J 58:1068–1082

Zhang H, Zhu JK (2012) Active DNA demethylation in plants and animals. Cold Spring Harb Symp Quant Biol 77:161–173

Zhang JX, Nguyen HT, Blum A (1999) Genetic analysis of osmotic adjustment in crop plants. J Exp Bot 50:291–302

Zhao B, Liang R, Ge L, Li W, Xiao H, Lin H et al (2007) Identification of drought-induced microRNAs in rice. Biochem Biophys Res Commun 354:585–590

Zheng XG, Chen L, Li MS, Lou QJ, Xia H, Wang P et al (2013) Transgenerational variations in DNA methylation induced by drought stress in two rice varieties with distinguished difference to drought resistance. PLoS One 8, e80253

Zhou L, Liu Y, Liu Z, Kong D, Duan M, Luo L (2010) Genome-wide identification and analysis of drought-responsive microRNAs in *Oryza sativa*. J Exp Bot 61:4157–4168

Zhu JK (2008) Epigenome sequencing comes of age. Cell 133:395–397

Zong W, Zhong XC, You J, Xiong LZ (2013) Genome-wide profiling of histone H3K4-tri-methylation and gene expression in rice under drought stress. Plant Mol Biol 81:175–188

Genomics of Temperature Stress

Paula Andrea Martinez

Introduction

An intrinsic behavior of any organism is to gain adaptation mechanisms for survival, which are specially developed over changing conditions. This homeostatic process of self-regulation (to maintain the stability of the system under fluctuating conditions or stress) is not yet well understood, due to the complexity of the stresses involved. For instance, temperature variations affect all levels of biological adaptation (Hickey and Singer 2004). In plants, temperature stress impacts on a large-scale and with a fast response, inducing direct alterations in plant development, growth, reproduction and survival. The ecological significance of studying such a stressor resides in understanding how climatic patterns are influential to plant fitness.

Plants species are adapted to a broad range of temperatures. It is known that the impact of temperature stress is not a steady condition; it is variable and it might differ seasonally, regionally and even on a daily basis. Plants sense these temperature alterations through complex and interrelated temperature-sensitive devices, e.g. membrane fluidity, protein conformation, enzymatic activities, cytoskeleton depolymerization, and metabolic reactions (Ruelland and Zachowski 2010). The process of adaptation to changing stress conditions releases a chain of metabolic reactions, i.e. to adjust energy uptake, water use, photosynthesis, sugar levels, developmental processes, etc. In turn, these reactions allow the plant to function and survive.

Despite having great plant diversity in all biomes of the world (major geographic regions that contain distinctive ecological communities), most commonly known plants have a small range of temperatures in which they can thrive. In particular, for many commercial crops the nearly ideal growth temperature is around 23 °C daytime

P.A. Martinez, M.Sc. (✉)
School of Agriculture and Plant Sciences, University of Queensland,
83 Harley Teakle Building, Brisbane, QLD 4072, Australia
e-mail: paula.martinez@uq.net.au

D. Edwards, J. Batley (eds.), *Plant Genomics and Climate Change*,
DOI 10.1007/978-1-4939-3536-9_6

and 15 °C at night (Gusta 2012). Thereafter, crop yields will face deleterious effects by the ever more frequent range of climatic extremes predicted for the upcoming decades (Easterling et al. 2000; Kharin et al. 2007; Coumou and Rahmstorf 2012), particularly temperature increments (Ciais et al. 2005; Alexander et al. 2006; Barriopedro et al. 2011; New et al. 2011). Encompassing all possible alterations in the climatic conditions, rigorous strategies for water use efficiency in croplands will need to be planned for the upcoming years. Hence, agriculture needs to find sustainable ways to produce the necessary amounts of food, whilst facing scarcity of water and temperature increment.

It is important to understand the broad range of environmental conditions which a plant needs to adapt, in order to manage regulation effectively. Such a complex set up presents challenges in the study of the physiological responses of temperature stress and ways to acquire tolerance. This is where the genomics perspective is best placed. In the last decade, the increase in genome sequencing projects from different plant species has greatly accelerated the impact of genomics on many areas of plant science and integrative biology (Bevan and Walsh 2005). For instance, forward and reverse genetics has been used to identify genes/pathways essential for acclimation to temperature stress (Hong et al. 2003; Shinozaki et al. 2003), structural genomics studies have been applied for linkage map construction, quantitative trait loci analysis (QTL) (Cattivelli et al. 2002; Saito et al. 2004; Xiao et al. 2011), genomic diversity analyses (Nevo 2001), genome wide association studies (Swindell 2006; Clarke and Zhu 2006; Huang et al. 2010), and marker-assisted selection (MAS) for stress tolerance (Abe et al. 2002; Toth et al. 2004; Francia et al. 2005). All these techniques generate valuable expertise for further analyzes, in both coding and non-coding regions associated with stress tolerance, as the basis of adaptation and speciation.

High Temperature Stress

Consequences of Heat-Stress

Of all the abiotic stress factors, temperature has a major impact on plant growth and reproduction. Heat stress is a complex stressor that comes in conjunction with other stressors, for instance lack of water availability, which promotes dehydration. Generally, heat stress is a condition of increased temperature to which the plant is not yet adapted to. At higher temperatures, stages of growth and development can be irreparably damaged. The degree of the heat stress impact can be assessed depending on the duration of the heat wave, occurrence (frequency), and the timing (seasonality). Also the heat-threshold level varies considerably at different developmental stages (Wahid et al. 2007). Thus, this complex stressor affects the equilibrium of plant metabolism from divergent points. Immediate responses to heat stress involve cellular, physiological and developmental changes.

Ideal temperatures for plant development vary according to the species, geographical areas and diverse ecosystems across the globe. The key factors to favor plant growth are photosynthesis, respiration and transpiration, which are affected by changes in daytime temperature. Respiration increases with the increase in temperature, when this increase overlaps the rates of photosynthetic build-up plant growth is eventually halted and sugars are rapidly synthesized to restore energy and to subsist. When heat increases, transpiration also increases and the plant starts to lose water to cool down. If the temperature is very high the plant will commence dehydration, as a consequence of water loss. Hence, the complete growth cycle of the plant is affected and needs physiological readjustments to survive the extreme changes in temperature.

Seed development may be inhibited by high temperatures resulting in diminished seed set and reduced seed weight (Blake 1935). Flowering is also adversely affected, as heat might cause flower drop or pod decay (Angadi et al. 2000; Larkindale and Vierling 2008), particularly in pulse legumes (Guilioni et al. 1997). An example, from a study of heat stress responses in rice by Shah et al. (2011), proved that spikelet sterility is induced by an altered hormonal balance in the floret, alterations in sugar biosynthetic enzymes and reduction of pollen grains ability to swell. Thus, decreasing rice yield under high temperature stress (Shah et al. 2011). In the long run, many crop plants are all likely to be negatively affected by warmer conditions in the coming years (Jarvis et al. 2010).

Direct damage due to high temperatures include protein denaturation, induced increment of lipids membrane fluidity, also called loss of membrane integrity (Wahid et al. 2007). Smertenko et al. (1997) described how the organization of microtubules is affected by heat stress by splitting and/or elongating of spindles during mitosis. Metabolic reactions also include reduced ion flux and production of toxic compounds and reactive oxygen species (ROS) (Mittler et al. 2004). Enhanced expression of a variety of heat shock proteins (Hsps) (Parsell and Lindquist 1993) and other stress-related proteins has been widely studied as a major component of plant responses to heat stress.

When a plant is exposed to a significant temperature increment, the first constrained resource will be water. The reduction of water produces desiccation. Some plants exhibit desiccation tolerance by dormancy, and the possibility to gain this state comes in relation to the ability to produce sugars (e.g. sucrose, raffinose and trehalose). Sugars are involved in the establishment of cellular protection prior to desiccation (Cushman and Bohnert 2000). In this case plants will survive in a suspended metabolic state until re-hydration occurs. Sugars had been shown to provide protective effects in the chloroplast membranes against heat stress (Santarius 1973). For example, studies in flowering plants revealed that highly soluble compounds (osmoprotectants), like sucrose and other sugars, raise osmotic pressure in the cytoplasm, and can also stabilize proteins and membranes when salt levels or temperatures are unfavorable (McNeil et al. 1999; Rizhsky et al. 2004).

Crops with an optimum daytime temperature of around 25 °C (Friend and Helson 1976), will be adversely impacted if their thermoperiod is affected. By the end of the twenty-first century, temperature is predicted to increase by 2–4 °C (New et al. 2011; IPCC 2007) and this increase at the molecular level can have many detrimental effects which will result in reduced yield.

Thermotolerance and Acclimation to Heat-Stress

Plants must evolve metabolic and structural adjustments in order to become tolerant to stress. Thermotolerance is gained after a prior exposure to a number of moderate stress treatments (in this case heat), also called conditioning pretreatment (Lindquist 1986). Thus during hotter days/seasons, plants may be in the continuous process of acquiring thermotolerance to maintain optimal growth (Hong and Vierling 2000). For example, the exposure to heat extremes in the desert might explain why these wild plants are far more resilient in hotter and drier conditions, than plants from temperate zones where heat extremes are less frequently observed.

Plants need to adjust their metabolism at the cellular level in response to a stress stimulus; this adjustment will act as a defense by expressing or suppressing genes that produce specific proteins (Yamanouchi et al. 2002). Hong and Vierling (2000) affirmed that there was insufficient understanding of how organisms acquire thermotolerance, owned to the lack of systematic genetic studies that analyzed the components involved as a whole. However, later studies using whole-genome analysis elucidated new findings in relation to regulatory functions in stress tolerance, (Matsui et al. 2008; Zeller et al. 2009; Urano et al. 2010).

Many proposed heat-stress tolerance mechanisms in plants have been studied in past decades. First, membrane fluidity is affected and the need to regulate membrane stability is important (Horváth et al. 1998). A decrease in the degree of fatty acid desaturation helps maintain the stability of the membrane and provides tolerance to heat stress (Graham and Patterson 1982; Grover et al. 2000). Second, signaling in the cytoplasm produced by reactive oxygen species (ROS) and scavenging of ROS is a tolerance mechanism (Bowler et al. 1992) which comes in combination with the production of antioxidants, accumulation and adjustment of compatible solutes for osmotic adjustments (Sung et al. 2003). Third, signaling pathways, the induction of mitogen-activated protein kinase (MAPK) and calcium-dependent protein kinase (CDPK) cascades are reactions of stress, as well as chaperone signaling and transcriptional activation factors (Knight and Knight 2001). Moreover, when heat is extreme, it induces denaturation of proteins and can cause cell death; the expression of heat shock proteins (Hsps) is known to be an important adaptative strategy relevant to this (Feder and Hofmann 1999). Many studies have involved the beneficial aspects conferred by the HSPs in heat tolerance, for example: photosynthesis (Al-Khatib and Paulsen 1990; Pastenes and Horton 1996), effective use of water and nutrients (Ackerly et al. 2000), and membrane stability (Frova and Gorla 1993). All these mechanisms regulated at the molecular level, enable plants to survive under heat stress (Wahid et al. 2007).

Larkindale and Vierling (2008) used whole-genome microarrays to analyze the transcription of thermotolerance related genes, with different acclimation treatments, in *Arabidopsis thaliana*. A gradual acclimation from 22 to 45 °C over 6 h showed better results than the more frequently used step-wise acclimation treatment where the temperature is increased, recovered to normal, and then increased once more. The results supported that gradual acclimation showed more effective protective responses to heat stress than the typical acclimation treatment. Using cluster analysis, Gusta (2012) identified major groups of transcripts and approximately one third of the up-regulated transcripts were heat shock proteins (Hsps) and molecular chaperones; in agreement with Larkindale and Vierling (2008) who demonstrated that Hsps alone are not sufficient for thermotolerance. Another third of the up-regulated transcripts were involved in plant recovery (specialized proteins accomplishing metabolic functions) (Gusta 2012). The other up-regulated transcripts are associated with photosynthesis, transcription/translation and protein degradation, abscisic acid response element (ABRE), phytohormone abscisic acid response, DNA dehydration-responsive elements (DRE) and anti-programed cell death (PCD) transcripts (Gusta 2012). Larkindale and Vierling (2008) found association between the down-regulated genes and other stresses, such as PCD and disease resistance genes. Complexgene pathways interaction show responses to a combination of concurrent stresses.

Low Temperature Stress

Consequences of Low Temperature Exposure

Growth, yield and distribution of plants are easily influenced by low temperature stress (Xin 2000). Plants better adapted to cold are those which are regularly exposed to low temperatures. This is because plants require days, or even weeks, to acclimate to cold and drought stress, in contrast with days or hours to acclimate to heat stress (Thomashow 1999). Some species are able to acclimate to stresses better than others, whereas cultivated species often have a limited capacity to do so. For instance the most intensive wheat cultivation occurs in temperate latitudes of both hemispheres (Leff et al. 2004). A similar situation is presented for other economically important crops (e.g. cereals, tubers and oil-bearing crops) which are better adapted to grow in temperate areas of the world (Wahid et al. 2007), and this group of crops are severely affected when a steep drop of temperature occurs.

The degree of injury caused by low temperature stress is defined by the duration, the timing and the degree of the thermal shock (an abrupt drop of temperature). Additionally, low temperatures are accompanied by short daytime and low sunlight radiation (Źróbek-Sokolnik et al. 2012), making low temperature stress a combination of environmental factors and a complex attributes. Although some plants require a period of low temperature exposure in the vernalization state (a required low temperature period to induce flowering) (Sheldon et al. 2000) this is not considered a stressor. However, if thermal shock occurs during spring, then it becomes a stressor. In fact, there are two types of cold stresses depending on the degree of the low-temperature stress. This could be either a chilling stress, when the temperature drops but is still above 0 °C, or a freezing stress when the temperature drops below 0 °C (Larcher and Bauer 1981). There are plants resilient to one or both of these stresses and they have different characteristic mechanisms to resist the lethal consequences of low temperature stress (Źróbek-Sokolnik et al. 2012).

In general, the initial impact of low temperatures is slowed photosynthesis which results in reduced growth and lower yields. In particular, chilling and freezing stress affect the entire internal environment of the cell. For instance, cold stress incites changes in the permeability of the membranes and other membrane dysfunctions (Źróbek-Sokolnik et al. 2012). At this point, called the unsaturation state, the membrane fatty acids react as a complement to membrane rigidification in an attempt to give protection against cold stress. Then, membrane lipids try to acquire a higher degree of unsaturation (Grover et al. 2000; Źróbek-Sokolnik et al. 2012; Ruelland et al. 2009). As previously mentioned the generation of reactive oxygen species (ROS) accompanied by oxidative stress are typical responses associated with stress, especially cold stress (Mittler et al. 2004; Blokhina et al. 2003), which can cause detrimental effects in important biomolecules of the cell. Additionally, chilling stress is one of the causes of increases in the concentration of free calcium ions in the cytosol (Minorsky 1989) which affects the intercellular cytoplasmic channels, called plasmodesmata, and limit the transport of water, nutrients and other molecules (Holdaway-Clarke et al. 2000). Moreover, cold affects the stability of proteins or protein complexes which is followed up by a protein restructuring. Consequently, this disturbs the metabolic regulations of the cell (Ruelland et al. 2009) and the thermodynamic equilibrium of the plant is disrupted.

Tolerance and Acclimation to Low Temperature Stress

The gain of stress tolerance in plants is a very complex mechanism, given that a single stress-tolerance gene has multiple effects on plant-stress tolerance (Gusta 2012). Additionally, plant organs also have diverse sensitivity to cold stress (Swindell 2006; Rorat 2006), and the stress responses might vary at different time points of developmental stages in plants (Sparks et al. 2007). Hence, the mechanisms of acclimation to low temperature stress are various; however these require the plant to be pre-exposed to periods of cold, to be able to trigger adequate responses to subsequent exposures.

If the temperatures are gradually lowered the plant will gain hardness, which is a usual way to acquire tolerance to cold stress. At this stage, signaling cascades and transcriptional control take place. Zhang et al. (2004), stated that plant adaptation to cold and drought is to a great extent dependent on transcriptional factors regulation. Thanks to transcriptomic studies, the expression of some transcription factors (TFs) which confer chilling and freezing tolerance to plants, had been revealed. For instance, cold stress initiates the expression of ABA (abscisic acid) growth regulators, CBF (C-repeat Binding Factor) and DREB (Dehydration responsive element binding), which are important transcription factors that activate many downstream genes to help stand cold temperatures (Kasuga et al. 1999; Chinnusamy et al. 2003; Yamaguchi-Shinozaki and Shinozaki 2006).

Additional ways to tolerate cold stress are more related to membrane stability; an increase in the degree of fatty acid unsaturation helps maintain the stability of the membrane and give tolerance to cold (Graham and Patterson 1982). For example the genetic manipulation of fatty acid desaturation was proved to be beneficial to gain tolerance to cold stress in engineered tobacco plants (Murata et al. 1992). Murata et al. (1992) also stated that other factors will play significant roles in the acclimation of transgenic plants and this example only proves that fatty acids are an important contributor to cold tolerance. Another mechanism to gain low temperature tolerance is to enter a dormant stage. At this stage, plants stop metabolic processes of growth (Liu et al. 1998), until the period of low temperature concludes. In dormancy, plants reduce their energy use and start to accumulate sugars as osmoprotectant (Źróbek-Sokolnik et al. 2012). For example, a comparative metabolite analysis in *Arabidopsis* by Urano et al. (2010) indicated that heat and cold stress regulation might also depend on the accumulation of other temperature-regulator metabolites, such as sucrose, proline, monosaccharides (glucose, fructose), raffinose, galactinol and myo-inositol.

Transcriptional profiling is used to analyze RNA-binding proteins which modulate the function of other regulatory factors. A good example of this approach is the whole genome analysis of RNA binding proteins of *A. thaliana* (Lorković 2009). In this analysis the authors confirmed the expression of glycine-rich RNA-binding proteins (GR-RBPs) and small RNA-binding proteins (S-RBPs) which stabilize the cell wall after stress stimuli (Didierjean et al. 1992; Dunn et al. 1996). Essentially, the conformation of RNA secondary structures is affected by cold stress, and consequently gene and protein expression are altered (Ruelland et al. 2009) enabling enzymatic equilibrium to be re-established (Ruelland and Zachowski 2010).

The aforementioned studies clearly show the complexity of responses to cold stress. Experiments led by Gusta (2012) demonstrated that an increase in plant frost tolerance of 1 °C is considered to be significant; due to the highly complex nature of plant adaptation to low temperature (Sanghera et al. 2011). Given these points, there are many ongoing approaches to the development of new traits for tolerance and to gain a more complete understanding of cold acclimation. Thanks to the new technological resources many valuable improvements for sustainable food production can be attained.

References

Abe F, Saito K, Miura K, Toriyama K (2002) A single nucleotide polymorphism in the alternative oxidase gene among rice varieties differing in low temperature tolerance. FEBS Lett 527(1):181–185

Ackerly DD, Dudley SA, Sultan SE, Schmitt J, Coleman JS, Linder CR et al (2000) The evolution of plant ecophysiological traits: recent advances and future directions: new research addresses natural selection, genetic constraints, and the adaptive evolution of plant ecophysiological traits. BioScience 50(11):979–995

Alexander LV, Zhang X, Peterson TC, Caesar J, Gleason B, Klein Tank AMG et al (2006) Global observed changes in daily climate extremes of temperature and precipitation. J Geophys Res Atmos 111:D5

Al-Khatib K, Paulsen GM (1990) Photosynthesis and productivity during high-temperature stress of wheat genotypes from major world regions. Crop Sci 30(5):1127–1132

Angadi SV, Cutforth HW, Miller PR, McConkey BG, Entz MH, Brandt SA et al (2000) Response of three Brassica species to high temperature stress during reproductive growth. Can J Plant Sci 80(4):693–701

Barriopedro D, Fischer EM, Luterbacher J, Trigo RM, García-Herrera R (2011) The hot summer of 2010: redrawing the temperature record map of Europe. Science 332(6026):220–224

Bevan M, Walsh S (2005) The Arabidopsis genome: a foundation for plant research. Genome Res 15(12):1632–1642

Blake AK (1935) Viability and germination of seeds and early life history of prairie plants. Ecol monograph 5(4):408–460

Blokhina O, Virolainen E, Fagerstedt KV (2003) Antioxidants, oxidative damage and oxygen deprivation stress: a review. Ann Bot 91(2):179–194

Bowler C, Montagu M, Inze D (1992) Superoxide dismutase and stress tolerance. Annu Rev Plant Biol 43(1):83–116

Cattivelli L, Baldi P, Crosatti C, Di Fonzo N, Faccioli P, Grossi M et al (2002) Chromosome regions and stress-related sequences involved in resistance to abiotic stress in Triticeae. Plant Mol Biol 48(5-6):649–665

Chinnusamy V, Ohta M, Kanrar S, Lee B-h, Hong X, Agarwal M et al (2003) ICE1: a regulator of cold-induced transcriptome and freezing tolerance in Arabidopsis. Genes Dev 17(8):1043–1054

Ciais P, Reichstein M, Viovy N, Granier A, Ogée J, Allard V et al (2005) Europe-wide reduction in primary productivity caused by the heat and drought in 2003. Nature 437(7058):529–533

Clarke JD, Zhu T (2006) Microarray analysis of the transcriptome as a stepping stone towards understanding biological systems: practical considerations and perspectives. Plant J 45(4):630–650

Coumou D, Rahmstorf S (2012) A decade of weather extremes. Nat Clim Change 2(7):491–496

Cushman JC, Bohnert HJ (2000) Genomic approaches to plant stress tolerance. Curr Opin Plant Biol 3(2):117–124

Didierjean L, Frendo P, Burkard G (1992) Stress responses in maize: sequence analysis of cDNAs encoding glycine-rich proteins. Plant Mol Biol 18(4):847–849

Dunn MA, Brown K, Lightowlers R, Hughes MA (1996) A low-temperature-responsive gene from barley encodes a protein with single-stranded nucleic acid-binding activity which is phosphorylated in vitro. Plant Mol Biol 30(5):947–959

Easterling DR, Meehl GA, Parmesan C, Changnon SA, Karl TR, Mearns LO (2000) Climate extremes: observations, modeling, and impacts. Science 289(5487):2068–2074

Feder ME, Hofmann GE (1999) Heat-shock proteins, molecular chaperones, and the stress response: evolutionary and ecological physiology. Annu Rev Physiol 61(1):243–282

Francia E, Tacconi G, Crosatti C, Barabaschi D, Bulgarelli D, Dall'Aglio E et al (2005) Marker assisted selection in crop plants. Plant Cell Tiss Org Cult 82(3):317–342

Friend DJC, Helson VA (1976) Thermoperiodic effects on the growth and photosynthesis of wheat and other crop plants. Bot Gaz 137:75–84

Frova C, Gorla MS (1993) Quantitative expression of maize HSPs: genetic dissection and association with thermotolerance. Theor Appl Genet 86(2-3):213–220

Graham D, Patterson BD (1982) Responses of plants to low, nonfreezing temperatures: proteins, metabolism, and acclimation. Annu Rev Plant Physiol 33(1):347–372

Grover A, Agarwal M, Katiyar-Agarwal S, Sahi C, Agarwal S (2000) Production of high temperature tolerant transgenic plants through manipulation of membrane lipids. Curr Sci 79(5): 557–559

Guilioni L, Wery J, Tardieu F (1997) Heat stress-induced abortion of buds and flowers in pea: is sensitivity linked to organ age or to relations between reproductive organs? Ann Bot 80(2):159–168
Gusta L (2012) Abiotic stresses and agricultural sustainability. J Crop Im 26(3):415–427
Hickey D, Singer G (2004) Genomic and proteomic adaptations to growth at high temperature. Genome Biol 5(10):117
Holdaway-Clarke TL, Walker NA, Hepler PK, Overall RL (2000) Physiological elevations in cytoplasmic free calcium by cold or ion injection result in transient closure of higher plant plasmodesmata. Planta 210(2):329–335
Hong S-W, Vierling E (2000) Mutants of *Arabidopsis thaliana* defective in the acquisition of tolerance to high temperature stress. Proc Natl Acad Sci 97(8):4392–4397
Hong S-W, Lee U, Vierling E (2003) Arabidopsis hot mutants define multiple functions required for acclimation to high temperatures. Plant Physiol 132(2):757–767
Horváth I, Glatz A, Varvasovszki V, Török Z, Páli T, Balogh G et al (1998) Membrane physical state controls the signaling mechanism of the heat shock response in Synechocystis PCC 6803: identification of hsp17 as a "fluidity gene". Proc Natl Acad Sci 95(7):3513–3518
Huang X, Wei X, Sang T, Zhao Q, Feng Q, Zhao Y et al (2010) Genome-wide association studies of 14 agronomic traits in rice landraces. Nat Genet 42(11):961–967
IPCC (2007) Climate change 2007: synthesis report. In: Pachauri RK, Reisinger A (eds) Contribution of working groups I, II and III to the fourth assessment report of the Intergovernmental Panel on Climate Change. IPCC, Geneva, p 104
Jarvis A, Upadhyaya HD, Gowda CLL, Agrawal PK, Fujisaka S, Anderson B (2010) Climate change and its effect on conservation and use of plant genetic resources for food and agriculture and associated biodiversity for food security. Food and Agriculture Organisation of the United Nations (FAO), FAO Thematic Background Study Rome, Rome
Kasuga M, Liu Q, Miura S, Yamaguchi-Shinozaki K, Shinozaki K (1999) Improving plant drought, salt, and freezing tolerance by gene transfer of a single stress-inducible transcription factor. Nat Biotechnol 17(3):287–291
Kharin VV, Zwiers FW, Zhang X, Hegerl GC (2007) Changes in temperature and precipitation extremes in the IPCC ensemble of global coupled model simulations. J Climate 20(8):1419–1444
Knight H, Knight MR (2001) Abiotic stress signalling pathways: specificity and cross-talk. Trends Plant Sci 6(6):262–267
Larcher W, Bauer H (1981) Ecological significance of resistance to low temperature. Physiological plant ecology I. Springer, New York, NY, pp 403–437
Larkindale J, Vierling E (2008) Core genome responses involved in acclimation to high temperature. Plant Physiol 146(2):748–761
Leff B, Ramankutty N, Foley JA (2004) Geographic distribution of major crops across the world. Global Biogeochem Cycles 18(1):33 pages
Lindquist S (1986) The heat-shock response. Annu Rev Biochem 55(1):1151–1191
Liu Q, Kasuga M, Sakuma Y, Abe H, Miura S, Yamaguchi-Shinozaki K et al (1998) Two transcription factors, DREB1 and DREB2, with an EREBP/AP2 DNA binding domain separate two cellular signal transduction pathways in drought-and low-temperature-responsive gene expression, respectively, in Arabidopsis. Plant Cell 10(8):1391–1406
Lorković ZJ (2009) Role of plant RNA-binding proteins in development, stress response and genome organization. Trends Plant Sci 14(4):229–236
Matsui A, Ishida J, Morosawa T, Mochizuki Y, Kaminuma E, Endo TA et al (2008) Arabidopsis transcriptome analysis under drought, cold, high-salinity and ABA treatment conditions using a tiling array. Plant Cell Physiol 49(8):1135–1149
McNeil SD, Nuccio ML, Hanson AD (1999) Betaines and related osmoprotectants. Targets for metabolic engineering of stress resistance. Plant Physiol 120(4):945–949
Minorsky PV (1989) Temperature sensing by plants: a review and hypothesis. Plant Cell Environ 12(2):119–135

Mittler R, Vanderauwera S, Gollery M, Van Breusegem F (2004) Reactive oxygen gene network of plants. Trends Plant Sci 9(10):490–498
Murata N, Ishizaki-Nishizawa O, Higashi S, Hayashi H, Tasaka Y, Nishida I (1992) Genetically engineered alteration in the chilling sensitivity of plants. Nature 356(6371):710–713
Nevo E (2001) Evolution of genome–phenome diversity under environmental stress. Proc Natl Acad Sci 98(11):6233–6240
New M, Liverman D, Schroder H, Anderson K (2011) Four degrees and beyond: the potential for a global temperature increase of four degrees and its implications. Philos Trans A Math Phys Eng Sci 369(1934):6–19
Parsell DA, Lindquist S (1993) The function of heat-shock proteins in stress tolerance: degradation and reactivation of damaged proteins. Annu Rev Genet 27(1):437–496
Pastenes C, Horton P (1996) Effect of high temperature on photosynthesis in beans (I. Oxygen evolution and chlorophyll fluorescence). Plant Physiol 112(3):1245–1251
Rizhsky L, Liang H, Shuman J, Shulaev V, Davletova S, Mittler R (2004) When defense pathways collide. The response of Arabidopsis to a combination of drought and heat stress 1[w]. Plant Physiol 134(4):1683–1696
Rorat T (2006) Plant dehydrins—tissue location, structure and function. Cell Mol Biol Lett 11(4):536–556
Ruelland E, Zachowski A (2010) How plants sense temperature. Environ Exp Bot 69(3):225–232
Ruelland E, Vaultier M-N, Zachowski A, Hurry V, Jean-Claude Kader, Michel D (2009) Chapter 2 Cold signalling and cold acclimation in plants. Adv Bot Res 49:35–150
Saito K, Hayano-Saito Y, Maruyama-Funatsuki W, Sato Y, Kato A (2004) Physical mapping and putative candidate gene identification of a quantitative trait locus Ctb1 for cold tolerance at the booting stage of rice. Theor Appl Genet 109(3):515–522
Sanghera GS, Wani SH, Hussain W, Singh NB (2011) Engineering cold stress tolerance in crop plants. Curr Genomics 12(1):30
Santarius KA (1973) The protective effect of sugars on chloroplast membranes during temperature and water stress and its relationship to frost, desiccation and heat resistance. Planta 113(2):105–114
Shah F, Huang J, Cui K, Nie L, Shah T, Chen C et al (2011) Impact of high-temperature stress on rice plant and its traits related to tolerance. J Agric Sci 149:545–556
Sheldon CC, Rouse DT, Finnegan EJ, Peacock WJ, Dennis ES (2000) The molecular basis of vernalization: the central role of FLOWERING LOCUS C (FLC). Proc Natl Acad Sci 97(7):3753–3758
Shinozaki K, Yamaguchi-Shinozaki K, Seki M (2003) Regulatory network of gene expression in the drought and cold stress responses. Curr Opin Plant Biol 6(5):410–417
Smertenko A, DrÁBer P, Viklický V, Opatrný Z (1997) Heat stress affects the organization of microtubules and cell division in Nicotiana tabacum cells. Plant Cell Environ 20(12):1534–1542
Sparks T, Menzel A, Editor-in-Chief: Simon AL (2007) Plant phenology changes and climate change. Encyclopedia of biodiversity. Elsevier, New York, NY, pp. 1–7
Sung D-Y, Kaplan F, Lee K-J, Guy CL (2003) Acquired tolerance to temperature extremes. Trends Plant Sci 8(4):179–187
Swindell WR (2006) The association among gene expression responses to nine abiotic stress treatments in *Arabidopsis thaliana*. Genetics 174(4):1811–1824
Thomashow MF (1999) Plant cold acclimation: freezing tolerance genes and regulatory mechanisms. Annu Rev Plant Biol 50(1):571–599
Toth B, Francia E, Rizza F, Stanca A, Galiba G, Pecchioni N (2004) Development of PCR-based markers on chromosome 5H for assisted selection of frost-tolerant genotypes in barley. Mol Breed 14(3):265–273
Urano K, Kurihara Y, Seki M, Shinozaki K (2010) 'Omics' analyses of regulatory networks in plant abiotic stress responses. Curr Opin Plant Biol 13(2):132–138
Wahid A, Gelani S, Ashraf M, Foolad MR (2007) Heat tolerance in plants: an overview. Environ Exp Bot 61(3):199–223

Xiao Y, Pan Y, Luo L, Zhang G, Deng H, Dai L et al (2011) Quantitative trait loci associated with seed set under high temperature stress at the flowering stage in rice (Oryza sativa L.). Euphytica 178(3):331–338
Xin Z (2000) Cold comfort farm: the acclimation of plants to freezing temperatures. Plant Cell Environ 23(9):893–902
Yamaguchi-Shinozaki K, Shinozaki K (2006) Transcriptional regulatory networks in cellular responses and tolerance to dehydration and cold stresses. Annu Rev Plant Biol 57:781–803
Yamanouchi U, Yano M, Lin H, Ashikari M, Yamada K (2002) A rice spotted leaf gene, Spl7, encodes a heat stress transcription factor protein. Proc Natl Acad Sci 99(11):7530–7535
Zeller G, Henz SR, Widmer CK, Sachsenberg T, Rätsch G, Weigel D et al (2009) Stress-induced changes in the *Arabidopsis thaliana* transcriptome analyzed using whole-genome tiling arrays. Plant J 58(6):1068–1082
Zhang JZ, Creelman RA, Zhu J-K (2004) From laboratory to field. Using information from Arabidopsis to engineer salt, cold, and drought tolerance in crops. Plant Physiol 135(2):615–621
Źróbek-Sokolnik A, Ahmad P, Prasad MNV (2012) Temperature stress and responses of plants. Environmental adaptations and stress tolerance of plants in the era of climate change. Springer, New York, NY, pp 113–134

“Genes, Meet Gases”: The Role of Plant Nutrition and Genomics in Addressing Greenhouse Gas Emissions

Jennifer Ming-Suet Ng, Mei Han, Perrin H. Beatty, and Allen Good

Introduction

One of the most significant issues that humanity needs to address is our impact on global warming. In the past century, the burning of fossil fuels, deforestation, industrial practices, and agriculture have all contributed to climate change, resulting in the Earth’s average temperature increasing by 0.8 °C, with approximately 70 % of this increase since 1980, and with projections to increase another 1.1–6.4 °C over the coming century, depending on the model used (USGRP 2009) (Fig. 1). These changes will not occur consistently across the globe; instead, temperature increases will be more pronounced at the poles than at the equator. However, more relevant to agriculture will be the extent of temperature increases in the major agricultural areas. These temperature increases are believed to be driven by increasing concentrations of key greenhouse gases (GHGs) (IPCC 2007) and while there are always questions about the models used to predict temperature increases, the increasing levels of CO_2 and N_2O are not in dispute (Fig. 1). Clearly we need to rethink how we approach sustainability, and one key component will be rethinking how we approach agriculture.

Today, farming and agricultural are called upon to feed the global population of seven billion people. In the process of doing this, the agricultural industry has undergone significant changes over the past 50 years due to the ‘Green Revolution’. In addition to improved genotypes and crop varieties, irrigation was expanded, and industrial breakthroughs, such as the Haber-Bosch process, allowed farmers to utilize synthetic fertilizers, resulting in a doubling of crop yields (Fig. 2). However, as the global population rises to nine billion people by 2050, we need to improve on

J.M.-S. Ng • M. Han, Ph.D. • P.H. Beatty, Ph.D. • A. Good, B.Sc., M.Sc., Ph.D. (✉)
Department of Biological Sciences, University of Alberta,
77 University Campus, Edmonton, AB, Canada, T6G 2E9
e-mail: allen.good@ualberta.ca

D. Edwards, J. Batley (eds.), *Plant Genomics and Climate Change*,
DOI 10.1007/978-1-4939-3536-9_7

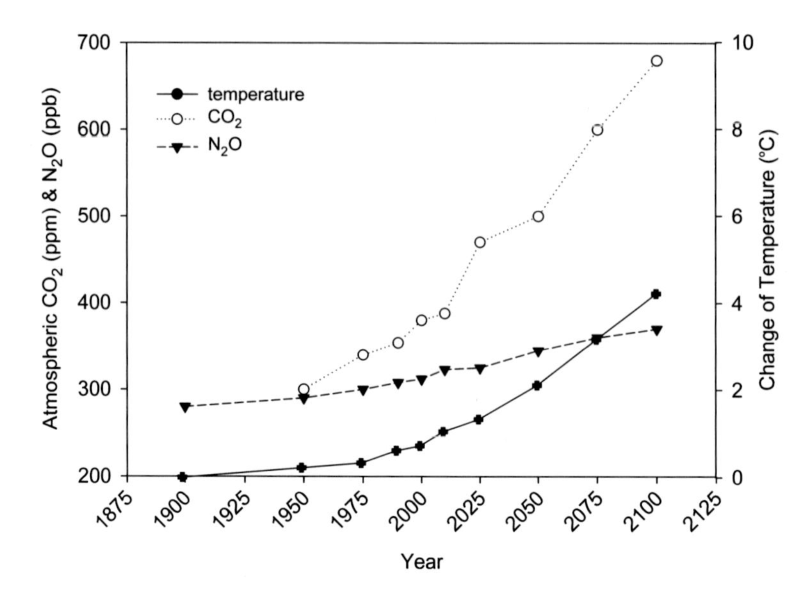

Fig. 1 Changes in key atmospheric GHGs between 1900 and 2010, with future projections until 2100. All future atmospheric projections for CO_2 value were based on scenarios B2 of Special Report on Emissions Scenarios (SRES) (Sitch et al. 2005), while N_2O emission level was adapted from Davidson et al. (2009) and (2012), Park et al. (2012) and the IPCC report (Park et al. 2012; IPCC 2001; Davidson et al. 2009, 2012; Ehhalt and Prather 2001) (ppm, parts per million; ppb, parts per billion). Changes of temperature were relative to the 1960–1979 average according to USGRP (2009)

the advances of the Green Revolution. Agriculture needs to redevelop and redesign methods to address the challenges we need to meet, and do so in a more sustainable manner. While it is a simple assumption that increasing the amount of applied fertilizer will continue to boost crop yields, excessive application of fertilizer does not increase yields in a linear relationship (Cui et al. 2008). On-farm and experimental plot studies show that increased application rates only increase N losses and are often ineffective at increasing yields past a certain point (e.g. China (Cui et al. 2010), for a review of the global perspective see (Fischer et al. 2009)). In fact, Cui et al. (2008) found that traditional producers N application rates of 325 kg ha^{-1} did not increase economic gains for the farmer over no additional N application, and resulted in losses of $146 $acre^{-1}$, compared to an N optimal application rate of only 130 kg ha^{-1}. Taken together, these findings show that the benefits associated with synthetic fertilizer applications to conventional crops have plateaued and will be unable to meet the needs of our growing population.

Within agriculture, the fields of agronomy, plant genetics and plant breeding have, and will, continue to play a role in maintaining or increasing yields while addressing issues surrounding global warming. Although yield gains have been significant over the last 60 years (Fig. 2), there is now overwhelming evidence that "business as usual" crop development will be insufficient to adapt crops over the

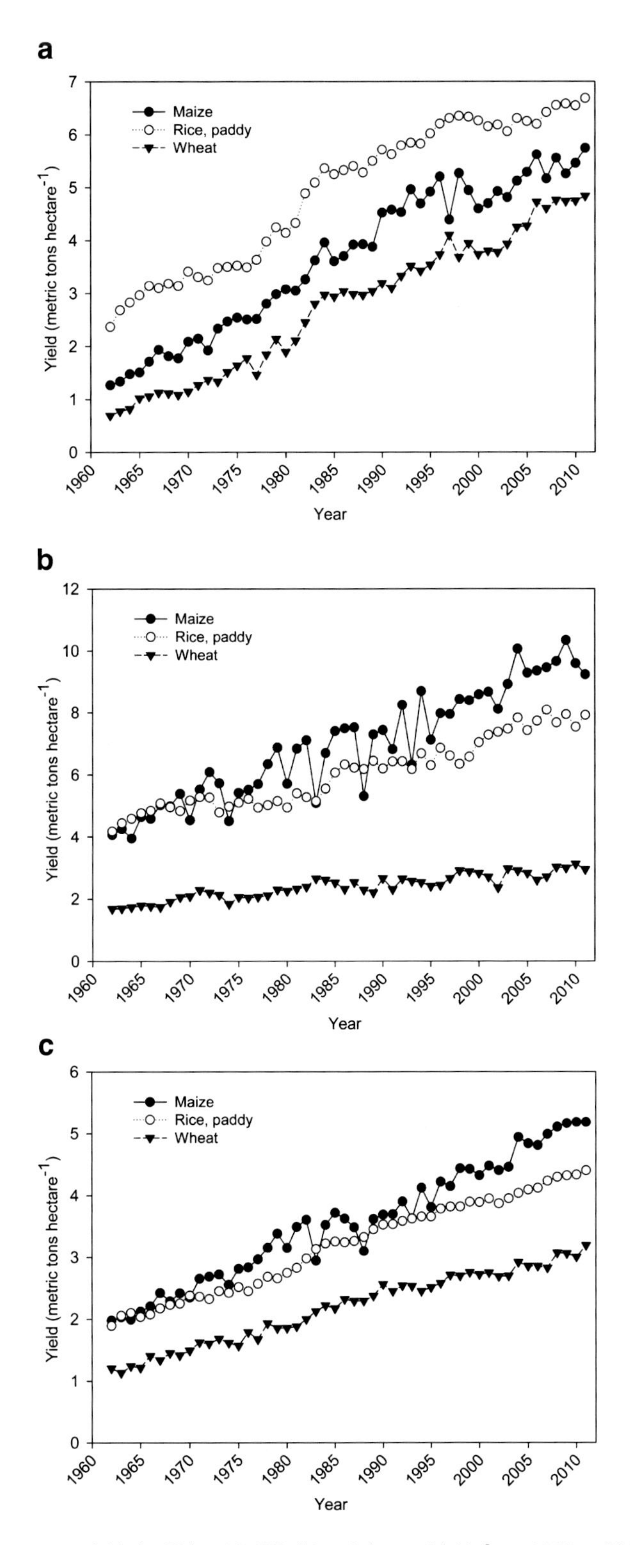

Fig. 2 Average crop yields in China (**a**), US (**b**) and the world (**c**) from 1962 to 2012. [Based on data from FAO (database) (http://www.fao.org/corp/statistics/en/)]

Table 1 Total N consumption, economic, and environmental costs for the US, China and the world

Year	Actual/predicted consumption (MMt N)	Value (US$B)	Proposed reduction (from 2005), %	Proposed consumption (MMt N)	Reduced use (MMt)*	N_2O emissions BAU $CO_{2\ eq}$	Reduced N_2O emissions $CO_{2\ eq}$	Value of reduced N_2O emissions (US$B)	Total environmental cost (US$B)
World									
1990	75.8	32.2				2871.3			
2005	100.6	80.0				3285.6			
2020	*110.7*	*108.5*	10.0	90.5	20.2	*3700.0*	6262.0	93.9	8.7–119.5
2050	*151.6*	*227.4*	20.0	80.5	71.1	*4528.7*	22,041.0	330.6	46.9–420.7
2100	*176.3*	*387.9*	10.0	90.5	85.8	*5909.9*	26,598.0	399.0	67.7–507.7
USA									
1990	9.5	4.1				397.6			
2005	14.5	11.5				356.1			
2020	*16.7*	*16.3*	5	13.8	2.9	432.1	899	13.4	1.2–17.2
2050	*19.9*	*24.3*	10	13.0	6.9	468.4	2139	32.0	6.9–40.8
2100	*23.5*	*35.2*	10	13.0	10.5	493.3	3255	48.8	8.3–62.1
China									
1990	18.6	7.1				304.8			
2005	34.8	27.6				393.7			
2020	42.4	42.0	10	31.3	11.1	756.6	3441	51.6	5.0–65.7
2050	54.5	66.5	20	27.8	26.7	1482.4	8277	124.1	27.1–158.0
2010	69	103.5	30	24.4	41.2	2692.1	12,772	191.6	32.5–243.8

Adapted from Good AG, Beatty PH. Fertilizing nature: a tragedy of excess in the commons. PLoS biology [Internet] 2011 [cited 2013 Mar 10];9 (8):e1001124. With permission from PLoS Biology.

*Difference between predicted and proposed

wide range of growing regions that will be required to meet expanding global agricultural demand (Long and Ort 2010). Current and future increases in temperature and associated droughts suggest that addressing the most significant and most urgent challenges for the adaptation of crops to global change will require all of our concerted efforts. Unless we address why agricultural yield gains seem to be plateauing, with the gains in recent years being much more modest than previous years, we will soon be unable to meet global food needs (Table 1).

The use of synthetic fertilizers has affected both the balanced composition and concentration of gases in our atmosphere. If high consumption of synthetic fertilizers continues, the significant increases in temperature and the unequal distribution of rainfall and water will continue to have destabilizing effects on food prices and economies and contribute to political unrest, such as what occurred during the Arab Spring of 2010 (Perez and Wire 2013). While the challenges may be greatest in the developing world and areas such as Sub-Saharan Africa (SSA), adapting crops in even the developed, high priority regions to meet the changing climatic conditions will require broad investment, the integration of new technologies with conventional selection-based breeding, and the coordinated involvement of public and private sectors of the agricultural enterprise (Long and Ort 2010).

What Are the Key Greenhouse Gases Emitted by Agriculture?

The key greenhouse gases include carbon dioxide (CO_2), methane (CH_4), nitrous oxide (N_2O), and fluorinated gases (F-gases such as hydrofluorocarbons, perfluorocarbons and sulfur hexafluoride), which in 2011 accounted for 84, 10, 4, and 1–2 % respectively of total U.S. GHG emissions (6702 MMt of CO_2 equivalents), and very similar levels in other regions (Table 2). Worldwide, we currently emit 33,376 MMt of GHGs, in CO_2 equivalents.

The IPCC uses the year 1750 as the key date for considering long-lived greenhouse gases (LLGHGs), defining this as the "pre-industrial" era. Historically, the levels of the key GHGs (CO_2, CH_4 and N_2O) in the atmosphere have remained relatively constant over much longer periods of time. Detailed measurements of LLGHGs in the atmosphere and of specific isotopes in ice core samples which have been made over time at a number of sites have provided a glimpse of the impact of industrialization. Antarctic ice-core records of carbon dioxide (CO_2) and the other key GHGs extending back 800,000 years at Dome C in the Antarctic, clearly shows that CO_2 levels have fluctuated between 180 and 280 ppm and N_2O levels between 220 and 300 ppb during that time period (Le Floch et al. 2013). However, the recent increases in both CO_2 and N_2O levels exceeds that range significantly, with the N_2O mixing ratio having increased by 20 % since 1750 (IPCC 2007; Park et al. 2012; MacFarling Meure et al. 2006).

While CO_2 levels are approximately 1000× higher than N_2O levels in the atmosphere, emission levels of CO_2 are also much higher (Fig. 1; Table 2). N_2O is 300× as potent as CO_2 and 12× as potent as methane (CH_4) per unit weight in its ability to

Table 2 GHG emission trends by greenhouse gas, in MMt CO_2 equivalents

Region	CO_2[a]	CH_4[a]	N_2O[a]	F-Gases[a,b]	Total[a]	Net emissions[c]	LUCF[d]	Year	Source
World	73 %	17 %	9 %	1–2 %	100 %	–	–	2005	1
	2,7555	6407	3286	548	37,797	–	–	2005	1
	2,3674	6020	3114	380	33,188	–	–	2000	1
	2,1091	5816	2871	239	30,018	–	–	1990	1
US	84 %	9 %	5 %	2 %	100 %			2011	2
	5613	587	357	145	6702	5797	–905	2011	2
	5883	613	430	121	7047	6213	–835	2000	3
	5004	644	398	94	6140	5067	–1073	1990	3
EU-27	82 %	9 %	7 %	2 %	100 %	–	–	2011	4
	3745	389	335	92	4553	4226	–327	2011	4
	4115	480	417	67	5069	4764	–305	2000	4
	4418	596	522	59	5586	5304	–282	1990	4
China	80 %	13 %	5 %	2 %	100 %	–	–	2005	5
	5976	933	394	165	7467	7046	–421	2005	5
	3073	720	264	NA	4057	3650	–407	1994	6

Data from WRI CAIT (2012), US-EPA (2013), UNFCCC (2003, 2013a, b, Cannell 2003)
[a]Not including GHGs from the LUCF sector
[b]Fluorinated gases: perfluorocarbons (PFCs), hydrofluorocarbons (HFCs), sulphur hexafluoride (SF_6)
[c]Including GHGs from the LULUCF sector
[d]LUCF (land-use change and forestry) emissions

trap heat. As well, the ozone depletion potential of N_2O is quantitatively similar to other ozone depleting substances such as hydrochlorofluorocarbons. Therefore, nitrous oxide must be recognized as both a potent greenhouse gas and a catalyst of ozone depletion.

There are two clear trends in GHGs that raise concerns. First, N_2O isotope composition studies have been able to distinguish between natural and anthropogenic sources of N_2O and have confirmed that the rise in atmospheric nitrous oxide levels is largely the result of an increased reliance on nitrogen-based fertilizers (Park et al. 2012). Second, there continues to be an increase in N_2O emissions, particularly from agriculture, with emissions in the US increasing by 17 % from 1990 to 2005 (US-EPA. United States and Environmental Protection 2013). Moreover, N_2O emissions are expected to grow by 5 % between 2005 and 2020 due to increased emissions from agriculture as food production accommodates the growing global population. Part of the reason for the growth in agricultural N emissions is the nonpoint source nature of the emissions. Unlike CO_2 and SO_4, which are often emitted from a single source (e.g. smokestack), and therefore easy to both measure and mitigate, N_2O, NOx and CH_4 emissions are nonpoint source pollutants that can accumulate in the biosphere from multiple source locations (e.g. farmer's fields).

In the US, agriculture contributed 6.3 % of total industrial greenhouse gas emissions in 2011. Methane emissions from enteric fermentation and manure management accounted for 23.6 and 8.9 % respectively, of total methane emissions from anthropogenic activities in 2011 (US-EPA 2013). Similarly, fertilizer application and crop-related activities alone represented 70.9 % of US nitrous oxide (N_2O) emissions for 2011. For the purpose of this chapter, the discussion on agriculture's impact on climate change will largely be limited to methane and nitrogen related emissions, and all emissions are reported in relation to a standard carbon dioxide unit (MMt in $CO_{2\ eq}$). Within agriculture, the biggest source of NOx emissions occurs in the form of N_2O and stems from soil management and the impact of synthetic nitrogen fertilizers applied to crops.

Clearly, stabilizing of the earth's surface temperature cannot be achieved by reducing CO_2 emissions alone, nitrogen gases and their role in climate change must be addressed as well. There have been a number of models that have looked at the reductions in GHG emissions, both in terms of CO_2 and the other key GHGs (Montzka et al. 2013). A 50–80 % cut in emissions by 2050 has been proposed as a means to limit GHG abundance below 500 ppm $CO_{2\text{-eq}}$, which is believed to be required to limit temperature increases to 2–2.4 °C (Montzka et al. 2013; Lucas et al. 2007). Emissions of several of the non-CO_2 GHGs must also be reduced, some of these reductions (10–15 % reduction in N_2O emissions) could occur with minimal net cost given the present technologies (Montzka et al. 2013; Shanahan et al. 2008). However, the key challenge for reducing N_2O emissions from agriculture is that they are non-point source emissions and are highly variable depending on many factors, such as weather and soil moisture (for a review of agricultural N_2O emissions (see Lucas et al. 2007).

Agriculture's Contributions to the Mitigation of GHGs and Climate Change

The impact of agriculture on climate changes is not entirely one-sided, while agricultural practices can result in increased anthropogenic greenhouse gas emissions, agriculture can also mitigate increasing GHG levels. For example, crops remove CO_2 from the atmosphere via photosynthesis and therefore agricultural soil can act as a carbon sink. Similarly, while the use of organic fertilizers releases N_2O, increasing levels of N entering marine ecosystem have been credited with providing much of the required N to allow the fixation of C (Montzka et al. 2013). These anthropogenic nitrogen additions have led to increased carbon uptake: approximately 10 % of oceanic CO_2 uptake is attributable to the atmospheric transport and deposition of anthropogenic nitrogen (Duce et al. 2008). Therefore, it is not always clear if and to what extent agriculture can be part of the solution.

Why Do N Emissions Occur? Asynchrony Between N Supply and Plant N Uptake

The global nitrogen cycle is notoriously "leaky" as N can be lost in number of ways, which have been well characterized by Gruber and Galloway (2008). Cereal crops tend to be very low in nitrogen use efficiency (NUE) and are therefore one of the major source for N loss to the environment. For example, Cui et al. (2008) found that N recovery efficiency under normal "farmer" conditions was only 18 %, but could be increased to 44 % with proper N management.

One approach that is often proposed is to genetically increase the N uptake and NUE of cereal crops. There have been a number of attempts to genetically manipulate the ability of plants to take up N more efficiently, which have been reviewed elsewhere (McAllister et al. 2012). N use by plants involves two main steps: uptake and utilization (Masclaux-Daubresse et al. 2010). N is most often taken up by plants as nitrate (NO_3^-; usually the most abundant form), ammonium (NH_4^+), and to a lesser extent, as organic compounds (Good et al. 2004; Miller et al. 2007; Rentsch et al. 2007; Näsholm et al. 2009).

Nitrate is taken up from the soil by roots using two main families of transporters; NRT1 and NRT2 (for a review see Miller et al. (2007)). The initial reduction of NO_3^- to nitrite (NO_2^-) occurs in the cytoplasm by nitrate reductase, with further reduction of NO_2^- to NH_4^+ occurring in the plastid/chloroplast by nitrite reductase (Masclaux-Daubresse et al. 2010). Ammonium is transferred to keto acids to synthesize various amino acids by glutamate synthase (GOGAT), glutamine synthetase (GS) and various aminotransferases (Good and Beatty 2011). Release of ammonium in leaf tissues due to remobilization of nutrients during senescence, as well as photorespiration in C_3 plants, requires that these tissues also have the ability to return N to the amino acid pool to be distributed as the plant requires (Liepman and Olsen 2003). The carbon skeletons utilized by these reactions are obtained from the tricarboxylic acid (TCA) cycle, making these reactions not only essential for N metabolism, but also important for C metabolism within the plant (Lawlor and Cornic 2002).

Once N has been taken up and assimilated, it is transported throughout the plant predominantly as NH_4^+ and amino acids for utilization and storage (Okumoto and Pilot 2011). Transported via the xylem, these N compounds are often distributed to mesophyll cells where they are either stored or utilized for carbon assimilation (Tegeder and Rentsch 2010). Chloroplast proteins make up approximately 80 % of the stored N in leaf tissues, with ribulose-1,5-bisphosphate carboxylase/oxygenase (Rubisco; carbon fixation enzyme) accounting for up to 50 % of the stored N in C_3 plants, and approximately 20 % of stored N in C_4 plants (Good and Beatty 2011; Kant et al. 2011). All of these steps potentially represent loss pathways and from an agricultural perspective, it is the ability for plants to effectively remobilize N into grain that is of critical importance to overall NUE, especially in cereal crops.

If ammonium and nitrate are not acquired by the plant, eventually they will be lost into the environment by leaching through the soil profile and into waterways, runoff of soil surfaces into waterways, volatized from the soil surface into the

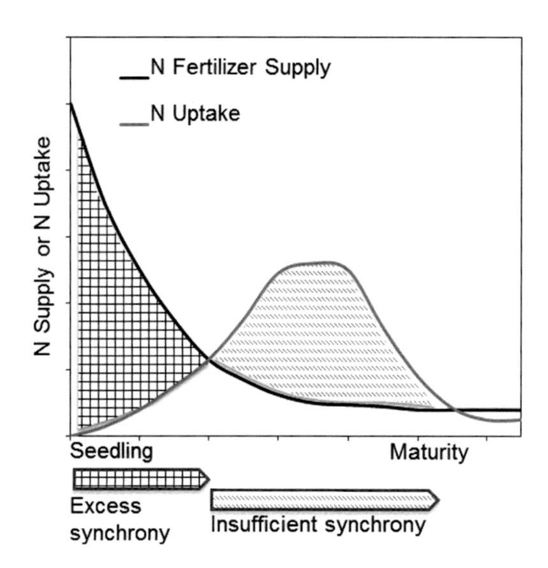

Fig. 3 The asynchrony between N supplied as synthetic N fertilizer at pre-plant or time of sowing and N uptake of the crops as they develop from seedlings to mature plants. Excess asynchrony occurs when there is more N fertilizer present in the soil than the plant can take up. Insufficient asynchrony is when there is less synthetic N fertilizer present in the soil than the crop needs at that time in its development

atmosphere and lost as NOx gases from soil microbial denitrification (Gruber and Galloway 2008; Crews and Peoples 2005). Nitrogen asynchrony occurs when the amount of N that is applied to the crop and the amount of N that is used by the crop are not in "synchrony" (Fig. 3). When crop demand for N and N supply (via fertilizer applications or organic matter mineralization) do not synchronize, that N has the potential to accumulate in soils and is then susceptible to various loss pathways (Goulding 2004; Peoples et al. 2004). Asynchrony can be divided into two types; excess and insufficient asynchrony. Excess asynchrony occurs when nutrient availability exceeds plant requirements, often because N fertilizer application coincides with pre-planting or seed sowing when plant N demand is non-existent or low, as in winter or early spring in temperate cropping systems. Insufficient asynchrony occurs when the plant N demand exceeds the N supply (Crews and Peoples 2005) (Fig. 3).

N asynchrony could be addressed in several ways. First, from an agronomic perspective, N fertilizer application (availability) could match plant N demand. This can be achieved in part by providing the nutrients at the critical time in the plant developmental cycle, when the plant requires a large N supply, as is often done in European countries after the adoption of the Nitrate Directive in 1987 (The EU Nitrates Directive. European Commission 2010; Frederiksen and Maenpaa 2007), which mandated the use of Best Nutrient Management Practices (BNMP). EU producers are now required to provide detailed N farm budgets before they can receive "Common Agricultural Policy" (CAP) subsidy payments (Good and Beatty 2011). Second, where possible, use information-intensive management approaches that can reduce GHG losses to the environment (Shanahan et al. 2008). Third, legumes could be used in rotations, for a source of green manure, which will be discussed later. Fourth, the inclusion of long-rooted perennials in cropping systems could also decrease the risk of N losses from the soil profile in future farming systems (Peoples et al. 2009).

The second feature that Fig. 3 illustrates is the concept that during the period of excess asynchrony, represented by the area where availability exceeds demand, much of the excess N is lost, either through GHG emissions or leaching. Regardless, there are two costs, first the direct cost of the fertilizer and second, the external costs associated with the environmental damage caused by this excess N. We have previously calculated these external costs as a minimum of 44 % of the direct fertilizer costs (Good and Beatty 2011), however other groups have suggested that these costs could be significantly higher (ENA 2011). For example, the European Nitrogen Assessment states that environmental costs in 2011, with the EU using 33.3 MMt of N fertilizer ranged from $26.3B USD to $191B USD, with the upper limit being well over 100 % of the fertilizer costs. Currently, prices do not reflect the full costs or benefits of consuming fertilizer or the benefit of a product that reduces these costs. These externalities are difficult to quantify and measure (particularly for emissions such as N_2O emissions from soils), however economic and environmental models need to be developed and used to understand these costs where excessive use of N occurs (Shanahan et al. 2008). Minimally we need to eliminate "negative green incentives" which often provide direct subsidies to farmers to use fertilizers. Some countries, such as Austria and Finland (Buttel 2003), have begun to implement "green taxes" (i.e. taxes on fertilizers and agrichemicals) and this trend needs to continue.

The Current Status of Plant Breeding Gains

One question that needs to be considered when looking at how to use plant genetics and genomics to address issues around climate change is to look where our successes have been in the past. Average yields for many key crops have increased steadily over the years. For example, average yield (per acre) has increased from 80 to over 180 bu/acre for irrigated maize in the US between 1960 and 2010 (while the average yields of non-irrigated (rain-fed) maize increased from 60 to 150 bu/acre during the same time period (Duvick 2005a) (Fig. 2). Similar increases in the average grain yield (per unit area) have been observed for rice, maize and wheat in China (Fig. 2a), where yields have increased by 90, 150 and 240 % respectively over the last four decades (Piao et al. 2010). Similarly, wheat and soybean yields have increased dramatically in the US, as have other key agricultural crops in the developed world (Fig. 2b). These significant gains in yield have allowed us to increase production, without dramatic increases in areas of cultivation. However, this situation may not continue. Many of the projected increases in yield are based on linear projections from our successes over the past 40–100 years (Miranowski and Rosburg 2011). A portion of these increases come from improvements in management practices, irrigation, fertilizers and herbicides, while the remainder has been from improved genetics. However, there is a general agreement that there are unlikely to be technologies that provide such significant gains in the future. One critical issue for addressing any crop improvements, regardless of whether they impact on GHG emissions is to understand the percent of total yield gains that can be attributed to genetics. It has been estimated that 50–60 % of the total yield gains can be attributed to the contribution of genetics during the past six or seven decades (Duvick 2005b).

While there is significant data which has supported our belief that increasing CO_2 levels will result in increased C capture and yield (IPCC 2007), other key changes associated with global warming are already affecting crops negatively, in particular heat and drought. While advances in overall yields have occurred, the question that is particularly relevant to plant nutrition is; to what extent can we select for improved nutrient use efficiency and to what extent have we made progress to date, in improving N uptake and/or N use efficiency.

Perhaps no crop has more data, although much of it held within companies, on the relationship between the genetic mechanisms involved in grain yield (GY) formation and its relationship with nitrogen and N uptake, than maize. In a review of the historical differences between maize varieties, Ciampitti and Vyn (2011) looked at studies on hybrids from 1940 to 1990—"Old Era" and studies from 1991 to 2011—"New Era." For the Old Era, maize grain yields (GY) averaged 6.8 Mg ha^{-1} with a total plant N uptake of 14.5 g m^{-2}, whereas for the New Era, maize GY averaged 9.5 Mg ha^{-1} and a total plant N uptake of 18.4 g m^{-2}. In maize and other crops, increases in overall yield and plant N uptake are often correlated. Ciampitti and Vyn (2011) also found that the New Era genotypes showed higher tolerance to N deficiency stress (higher GY when no N fertilizer was applied), and larger GY response per unit of N applied, relative to Old Era hybrids. In fact, while it is often a concern that breeding for higher yield may result in plants that do not produce well under low N conditions, this has not proved to be true in many cases. Hawkesford (2012) reviewed studies on breeding for increased NUE in crop plants and concluded that while there had been gains in NUE in cereal crops (rice, wheat, barley), these gains were the result of the selection for enhanced yield, under constant N conditions (something that has been noted by many researchers), rather than a conscious selection for enhanced NUE.

Genetics, Genomics and Addressing Climate Change

To begin, it is important to clarify what aspects of genomics we are specifically referring to. Genomics is often described as the large scale application of molecular biology and genetics in studying the role of different genes and gene combinations in determining a particular trait. However, for the purposes of this Chapter, we will not draw a distinction between more traditional genetic approaches, using markers and genetic crosses, versus transgenic approaches. For researchers and industry who are interested in addressing practical challenges, the question of how genomics can be used to address climate change has very practical considerations, in particular where they target their research efforts. However, it is also important to identify the opportunities for using genetic tools to mitigate GHGs in agriculture, since this will, to some degree determine the investment in the field.

Increasing Mitigation (CO_2)

Fundamentally the question is, “How can we use genetic tools to allow plants to more efficiently reduce the emissions of added N, or enhance the removal of GHGs (sequestration)?” Both of these approaches, one reducing the emissions of nitrous oxide, the other enhancing sequestration, could be significant in reducing the overall levels of GHGs (Bouwman et al. 2001).

The storage of C in plant matter and soils is one process that needs to be understood, in terms of where new technologies will need to focus. Many of the options for storing increasing amounts of C in the soil are agronomic and cannot be addressed by genetic technologies. Agricultural ecosystems hold large reserves of C (IPCC 2001), mostly in soil organic matter. Historically, these systems have lost significant amounts of C (Paustian and Cole 1998; Lal et al. 1999; Lal 2004) however some of this lost C can be recovered through improved management, thereby withdrawing atmospheric CO_2. Many studies worldwide have now shown that significant amounts of soil C can be stored, through a range of practices that are suited to local conditions (Lal 2004). Any practice that increases the photosynthetic input of C or slows the return of stored C via respiration will increase stored C, thereby ‘sequestering’ C or building C ‘sinks’.

However, it should be recognized that the role of N in C sequestration is going to be critical, as C and N metabolism are closely linked in plants and there is a good deal of experimental evidence that N limitation can often reduce the ability of an ecosystem (either agronomic or natural ecosystems) to sequester C. Much of this has not been fully integrated into our more recent models of climate change. For example, the recent IPCC report uses several models to predict increases in carbon sequestration with increasing CO_2 levels, however these do not take into account N limitation in many of these environments (IPCC 2007).

Agricultural lands also remove CH_4 from the atmosphere by oxidation, but this effect is small when compared with other GHG fluxes (Smith and Conen 2004). As mentioned earlier, significant amounts of vegetative C can also be stored in agroforestry systems or other perennial plantings on agricultural lands so forests should not be discounted (Albrecht and Kandji 2003). Since 1990, the carbon storage in US forests has increased and the total forested acreage has grown (Thomas et al. 2009), resulting in a significant C sequestration.

Finally, crops and residues from agricultural lands can be used as a source of fuel, either directly or after conversion to fuels such as ethanol or diesel (Schneider and McCarl 2003). These bioenergy feedstocks still release CO_2 upon combustion, but now the C is of recent atmospheric origin (via photosynthesis), rather than from fossil C. However, the economics and environmental benefits of these “green” technologies need to be looked at closely. In a recent review, Borak et al. (2013) showed that while an accurate, quantitative analysis of biological systems, accounting for both process and economics, would allow an analysis of how they impact on GHG emissions, a systematic methodology is currently lacking. Different disciplines, and even different research groups within a particular discipline, make different assumptions and use different comparators in the calculation of efficiencies and yields. Key physical data, including feedstock composition (particularly moisture content), seasonal yields, regional climatic conditions and year-over-year variability, are frequently not reported. This makes it difficult to come to objective conclusions about the impact of bioenergy crops on GHG emissions (Borak et al. 2013).

Carbon Dioxide: The Link Between N Fertilization and increased CO_2 Sequestration

One of the key ways in which crops plants are believed to be important in addressing climate change is through carbon (CO_2) fertilization. Since plants are known to fix CO_2, and the rate of CO_2 fixation is directly correlated with CO_2 concentrations, it has always been argued that increase CO_2 levels will result in enhanced sequestration. In principle, the removal of CO_2 by plants should compete with a number of the physical and chemical CO_2 sequestration strategies that may be able to reduce CO_2 emissions. In this way, the benefits of investing in research into either enhancing or minimally maintaining CO_2 fixation, in the face of increasing temperatures should become apparent, when compared to the significant investment in other types of carbon sequestration. The physical and chemical processes including; (1) physical separation and storage, such as direct injection of CO_2 into geological formations; (2) chemical absorption, based on chemical reactions of CO_2 with absorbent materials to capture and remove CO_2 from fossil fuel combustion (Yu et al. 2012), as well as the chemical processes to convert CO_2, NOx, and SOx emissions into fertilizers to enhance sequestration of CO_2 into the soil and subsoil layer (Lee and Lee 2003). In contrast, plants are not solely grown for CO_2 fixation, but rather this is a benefit from their natural biological processes. In theory, the biological CO_2 sequestration approach offers many potential advantages, including sequestrating relatively large volumes of carbon at comparatively low cost, producing usable products (food, feedstocks, biomass, bioenergy), and protecting soils, water resources and biodiversity. As such, it has been considered to be an economically feasible and environmentally sustainable technology for long-term CO_2 mitigation (Kumar et al. 2010; Stewart and Hessami 2005).

While there is a clear link between CO_2 sequestration and N fertilization, the relationship is a complex one and may not have been fully integrated into more recent models of climate change. Some simulation models assume that C can be sequestered, even when nutrient supplies are low (Cannell and Thornley 1998). Other models make perhaps a more realistic assumption that low N availability will preclude soil C storage. Van Groenigen et al. (2006) performed a meta-analysis of elemental interactions and how they limit carbon storage and found that elevated CO_2 causes accumulation of soil carbon only when N is added at levels that are well above typical atmospheric N inputs.

The Role of N Limitation and Genomics in Biomitigation

Biological CO_2 fixation or biomitigation involves the net removal of CO_2 from the atmosphere by plants, algae or other micro-organisms and the storage of CO_2 in vegetative biomass and soils. One way to mitigate increasing CO_2 levels would be to improve the efficiency of photosynthesis, the most common example in nature of biological CO_2 sequestration. It is estimated that approximately 258 billion tons of carbon dioxide are converted by photosynthesis organisms annually, and much of this is sequestered (Geider et al. 2001). More recently, microalgae have attracted a great deal of attention for CO_2 fixation and biofuel production because they can convert CO_2 (and nutrients) into biomass via photosynthesis at much higher rates than conventional biofuel crops can. It has been projected that annually around 54.9–67.7 tonnes of CO_2 can be sequestered from ponds of microalgae corresponding to annual dry weight biomass production rate of 30–37 tonnes per hectare (Lee and Lee 2003; Kumar et al. 2010; Stewart and Hessami 2005; Ho et al. 2011; IPCC 2005).

However, natural photosynthesis does not work with 100 % efficiency, with the process being limited by a key enzyme, ribulose 1,5-bisphosphate carboxylase/oxygenase (RuBisCO), in the CO_2 fixation step. In principle, genetically modified photosynthesis enzymes could provide a new strategy to further enhance CO_2 biological fixation. For example, it has been shown that genetically modified *E. coli* expressing genes encoding RuBisCO and a helper enzyme from photosynthetic bacteria are able to fix and convert CO_2 into consumable energy. Furthermore, when reconstructed with a more efficient RuBisCO variant, the *E. coli* strains grew faster, produced a larger quantity of the enzyme and led to faster assimilation of CO_2 gas (Parikh et al. 2006). Also, genetically engineered *Synechococcus elongatus* were able to use CO_2 as a direct substrate to produce isobutyraldehyde with increased productivity by overexpression of RuBisCO (Atsumi et al. 2009).

Efforts are underway by a number of groups on improving the efficiency of photosynthesis. For example, a project, titled "RIPE – Realizing Increased Photosynthetic Efficiency," was recently announced with a $25-million grant from the Bill & Melinda Gates Foundation to improve the photosynthetic properties of key food crops, including rice and cassava. This is in addition to a major initiative on improving the photosynthetic efficiency recently undertaken by the IRRI (International Rice Research Institute; http://www.gatesfoundation.org/Media-Center/Press-Releases/).

Reducing Emissions of the GHGs Methane and Nitrous Oxide

Anything that can be done to increase the ability of livestock to make efficient use of feeds will often suppress the amount of CH_4 produced (Clemens and Ahlgrimm 2001). This would also apply in a similar fashion to the CO_2 reductions one could gain from increasing the efficiency of converting maize into sugars for the production of ethanol. Before we begin to use genetics to improve crop NUE, and N losses, we need to also implement agronomic best management practices and reduce microbial denitrification in agricultural soils. Some examples of these practices include; improved matching of fertilizer applications to crop needs, expanded use of winter cover crops and nitrification inhibitors, improved drainage management, improved manure management, constructed wetlands and denitrification bioreactors, and riparian zone vegetation strips. All of these practical considerations do not require a genetic component, however they should be utilized alongside genetically improved crops to further reduce agricultural N_2O emissions.

It has been argued that using current technology, it is possible to improve agricultural N efficiency by up to 20–25 % (Park et al. 2012) and reduce N losses, including N_2O emissions, by 30–50 % (Cui et al. 2010; Good and Beatty 2011). The difficulty with using genetics to improve NUE is the challenge of phenotyping the trait and the cost of doing extensive field testing. Although there has been much promise of the newer phenotyping facilities, the reality is that field testing is still the only way to properly evaluate these traits, and developing field trials which include yield response curves is both challenging and costly (Langridge 2013).

Reducing the Need for Anthropogenic Fertilizers: Genomics and Legumes

One approach to reducing N emissions would be to use N fixing legumes much more frequently, both as feed and in rotations as green manure, than we currently do. The growth of crop plants (in both legume based and N fertilizer based systems) results in an increasing demand for available N in a curvilinear fashion as illustrated in Fig. 3. With legumes, the production of fixed N by symbiotic bacteria can occur as the plant develops so there is more synchrony between N supply and demand. In cereal crops, the asynchrony between excess N supply and low N demand at the early stages of crop growth results in much of the N loss in the biosphere. Studies that compare the fate of biologically fixed legume-N and fertilizer N suggest that a higher proportion of the legume N is retained in the soil, however provided that fertilizer N is used concurrent with best management practices, N losses tend not to differ greatly from either source (Drinkwater et al. 1998).

In applying genetics and genomics to the relationship between legumes and the nodulating bacteria, Rhizobia, several approaches could be considered. Research into the improvement of legume/bacterial interactions should be promoted. A practical approach to improving symbiosis would be to identify different genotypes from either plant or bacteria that promote enhanced nodulation rates and N fixation, although it should be noted that these are not always correlated.

Biological Nitrogen Fixation

Biological nitrogen fixation (BNF) is the process where inert, atmospheric N_2 gas is converted (fixed) to biologically reactive, or useful, ammonia. This reaction is only performed biologically in nature by diazotrophic bacteria. There are thousands of different species of diazotrophic bacteria and they fix N_2 using an enzyme complex called nitrogenase. Some diazotrophic bacteria form symbioses with plants, for example the nodulating bacteria-legume partnerships discussed previously. This is symbiotic because the plant trades photosynthetically derived carbon compounds for fixed N-ammonia from the bacteriods, to the benefit of both partners. The rate of symbiotic N fixation can be as high as 300 kg N ha^{-1} (Franche et al. 2008). Other diazotrophic bacterial species form associations with plants, for example researchers have discovered N fixing endophytic bacteria such as *Azospirillum* and *rhizobia* that grow endophytically within cereals (Saikia and Jain 2007; Yanni and Dazzo 2010). These associations appear to promote cereal growth, however it is unclear whether these plant growth-promoting bacteria (PGPB) give fixed N to their plant hosts (Stoltzfus et al. 1997). Other diazotrophic bacteria are called free living because they do not form relationships with plants but are instead found in aquatic or terrestrial environments with low fixed N levels.

There has been a recent insurgence of interest in research around BNF and cereal crops aimed at increasing the ability of cereal crops to acquire their fixed nitrogen either directly from diazotrophic bacteria or directly from the nitrogenase enzyme complex that catalyses the fixed N reaction (Godfray et al. 2010; Oldroyd et al. 2009; Den Herder et al. 2010). If we can engineer cereals to be able to fix their own N, we could greatly reduce, or perhaps even eliminate the need for N fertilizer applications. In addition, N supply would match cereal crop N demand allowing for synchrony between these two factors. With N synchrony and less fertilizer application, there would be much less N lost to the environment. Reducing or eliminating the need for N fertilizer application would reduce production costs for intensive crop farmers in developed countries and would allow smallholder farmers in developing countries to increase N supply to their crops.

There are three basic approaches that have been discussed in the literature recently to develop BNF cereals (Ladha and Reddy 2000; Reddy et al. 2002; Beatty et al. 2011). One approach would be to improve diazotrophic, endophytic bacteria-cereal plant associations (Yanni and Dazzo 2010; Stoltzfus et al. 1997). This method has the benefit that the interactions are already present in nature and these bacteria are known to promote plant growth. Farmers could include the enhanced endophytic N fixing bacteria with the seeds at planting time. There are biofertilizers (also called bioinoculants) that are on the market that use plant growth promoting bacteria. Some of these biofertilizers have diazotrophic bacteria in their formulation. Although, the efficacy of these biofertilizers to provide fixed N to crops is not currently well understood, they have shown the ability to improve the nutrient uptake and yield of crop plants (Mahdi et al. 2010). Enhancements to these associations would need to address such factors as; increasing the stability of the association and encouraging the bacterial partner to share it's fixed N with the plant. However, the limitations are that currently the low rate of N fixation from these associative bacteria may not be able to be increased enough to allow for a substantial reduction in N fertilizer application in intensive farming. Conversely, even modest increases in BNF capability within cereals would greatly increase crops yields in chronically N poor regions such as Sub-Saharan Africa.

Another approach to BNF cereals is to engineer cereal crop plants and nodulating diazotrophic bacteria to form symbiotic partnerships (Oldroyd et al. 2009; Ladha and Reddy 2000). BNF researchers have recently discovered that plants have the capacity to attract and form symbiotic partnerships with arbuscular mycorrhizal fungi and that many of the genes needed to do this have high similarity to the genes necessary for rhizobial-legume symbiosis (Maillet et al. 2011). This would suggest that some of the genes needed for symbiosis with diazotrophic bacteria are already present in cereals (Gherbi et al. 2008; Charpentier and Oldroyd 2010). Symbiotic N fixation rates can be comparable to the N fertilizer rates used in intensive farming, suggesting that if cereals could be engineered to allow for diazotrophic colonization and nodulation, they could have a high level of *in planta* sourced N to use in biomass and yield. The benefits of this approach to farmers would be that they could grow cereal crops like legumes where N-fertilizer applications to the growing crop could be greatly reduced or even eliminated just as long as the correct symbiotic bacterial partner was planted with the cereal crop. However, an understanding of the legume-rhizobial symbiosis process has increased greatly in recent years. Researchers at the John Innes Centre in UK have recently received a grant of over $9 million USD from the Bill & Melinda Gates Foundation (B&MGF) to study the feasibility of developing diazotrophic symbiosis with traditionally grown Sub-Saharan maize (www.foundationcenter.org/gpf/foodsecurity).

A third approach to engineering BNF cereals that would address many of the issues associated with excessive N applications, but would be just as challenging and as lengthy a research process as engineering symbiotic cereals would be to introduce the genes encoding nitrogenase directly into a plant (Beatty et al. 2011; Merrick and Dixon 1984). The approximately 12–20 bacterial genes essential for nitrogenase activity could be used to stably transform cereal crops so that they would be able to fix their own N using freely available N_2 gas from air. There are many challenges to this approach with one of the largest being that the nitrogenase enzyme is oxygen labile, so it would need to be stably introduced into a microaerobic environment in the plant to ensure functionality, such as an organelle (Merrick and Dixon 1984). Also the potential rate of biological N fixation for the plant is unknown, although it is possible that it could match the rate seen in legume symbiosis. Fundamentally, the key benefit of this approach to farmers would be that the plant itself can meet it's own N demand so there would be no need to introduce an N fixing partner at the time of sowing and N fertilizer application could be greatly reduced or eliminated.

Focusing Our Resources

While the use of genetics and genomics in plant development and breeding can be targeted to any trait, whether the trait can be easily measured, or whether the variability that exists is too great to allow a genetic dissection of the trait, has always been a critical problem. Therefore, in deciding where to invest scant resources, a number of factors come into play, including the following; (1) what is the trait that one wishes to measure? (2) What is the crop and is there value in the crop? (3) What is the value of the trait? (4) Where can genomics be effective in addressing issues of plant nutrition?

Choice of Crop

Realistically, the key crops which companies and most organizations are willing to invest in are maize, rice, wheat, rapeseed (*Brassica napus*), cotton and soybean. Other crops have had some investment in genetic resources (e.g. other members of the Brassicaceae (Parkin et al. 2010)), but taking these resources to the field and applying them in addressing issues of GHG emissions remains challenging.

The next question that needs to be addressed is; what is the value of the trait? If a trait is transgenic, then the cost of gaining regulatory approval in any jurisdiction is significant. Recent estimates of the cost of regulatory approval in the US, for a transgenic event are now between $25 M and $50 M, so that the trait must deliver sufficient return on investment to pay for both the initial research, the cost of evaluating the trait in the field and the cost of registration of the event. Clearly any Net Present Value calculation would indicate that the trait must deliver well in excess of $100 M in value to warrant the investment. While this may be true for some key crops, for many orphan crops (ones where there is a minor, but relatively insignificant investment in R&D), this is not the case.

Perhaps the most encouraging recent development has been the commitment of a number of NGOs, particularly the Bill & Melinda Gates Foundation, to funding very basic research into different approaches to address issues of chronic N shortages in SSA. In addition, they have been instrumental in evaluating the use of legumes to enhance biological N fixation through their N2Africa program. The N2Africa project is a 4-year international collaboration to research and implement improved nitrogen fixation for smallholder farmers growing legume crops in SSA. Due to high costs of fertilizer, only 10 % of smallholder farmers in Africa use fertilizers and as a result, these countries have poor grain yields due to low N availability which is significant since agriculture accounts for 19–41 % of the GDP in these nations (www.n2africa.org) . Because agriculture is the sole income for most smallholder farmers, improving legume productivity will increase farm incomes, thereby enhancing the quality of life for these people. The N2Africa project has the goal of raising average grain legume yields by 954 kg/ha in four legumes (groundnut, cowpea, soybean, and common bean), and increase average BNF capacities by 46 kg ha^{-1} (Njenga and Gurung 2011). If this is achieved, this will represent a significant increase in the level of biologically fixed N in SSA, by increasing the use of legumes to fix N, and the development of improved legume varieties and inoculants to maximize the efficiency of biological N fixation (Giller et al. 2009).

Conclusion

In order to successfully reduce N_2O emissions and N losses to the environment, we need to optimize the use of N fertilizers, for which there are initially a number of simple approaches. First, fertilizer use requirements need to be reassessed in virtually all agricultural systems, from an economic and environmental perspective. Second, economic and environmental models need to be integrated, particularly in those countries where excessive use of N occurs. As an example of this, Chen et al. (2011) recently suggested that an integrated soil-crop system could nearly double crop yields, with no increase in fertilizer usage (Chen et al. 2011). Much of the research has been conducted to look at fertilizer application rates, and different types of N fertilizers, however, the extent to which genotype can affect N emissions has yet to be determined. Finally, if the issues of N emissions are to be dramatically changed, we will need to rely much more heavily on biological nitrogen fixation with legumes and perhaps using novel genetic approaches, as has been recently suggested (Beatty et al. 2011).

Acknowledgements This research was funded in part by the Alberta Crop Industry Development Fund.

Glossary

BMNP Best management nutrient practices, or BMP, best management practices.

BNF Biological nitrogen fixation (BNF) is the process where inert, atmospheric N_2 gas is converted (fixed) to biologically reactive, or useful, ammonia. This reaction is only performed biologically in nature by diazotrophic bacteria.

Carbon dioxide (CO_2) The primary greenhouse gas emitted through human activities. In 2011, CO_2 accounted for 84 % of all anthropogenic U.S. greenhouse gas emissions.

CO_2 equivalent A metric measure used to compare the emissions from various greenhouse gases based upon their global warming potential (GWP). Carbon dioxide equivalents are commonly expressed as "million metric tons of carbon dioxide equivalents ($MMTCO_2Eq$)." The carbon dioxide equivalent for a gas is derived by multiplying the tons of the gas by the associated GWP. $MMTCO_2Eq$ = (million metric tons of a gas) × (GWP of the gas) (www.cga.gov)

Nitrous oxide (N_2O) Nitrous oxide (N_2O) accounted for about 5 % of all U.S. anthropogenic greenhouse gas emissions and is naturally present in the atmosphere as part of the Earth's nitrogen cycle (www.ega.gov)

Fluorinated hydrocarbons (FC) FC's include hydrofluorocarbons, perfluorocarbons, and sulfur hexafluoride and are synthetic, powerful greenhouse gases that are emitted from a variety of industrial processes. Fluorinated gases are sometimes used as substitutes for stratospheric ozone-depleting substances (www.ega.gov)

Methane (CH_4) Methane is a greenhouse gas with a GWP of 21 and is emitted during the production and transport of coal, natural gas, and oil. Methane emissions also result from livestock and other agricultural practices and by the decay of organic waste in municipal solid waste landfills.

NO_x A generic term for mono-nitrogen oxides NO and NO_2 (nitric oxide and nitrogen dioxide). NO_x should not be confused with nitrous oxide (N_2O), which is a greenhouse gas and has many uses as an oxidizer, an anesthetic, and a food additive.

NO_y Reactive (odd nitrogen) is defined as the sum of NO_x plus the compounds produced from the oxidation of NO_x which include nitric acid.

PGPB Plant growth-promoting bacteria, these may or may not be diazotrophic.

LUCF The term LUCF refers to (Land Use Change Forestry) provides an assessment of the net greenhouse gas fluxes (emissions of greenhouse gases, and removal of C from the atmosphere) resulting from the uses of land types and forests as well as those associated with land-use change.

IPCC The **Intergovernmental Panel on Climate Change** (**IPCC**) is a scientific intergovernmental body, first established in 1988 the World Meteorological Organization (WMO) and the United Nations Environment Programme (UNEP), and later endorsed by the United Nations General Assembly through Resolution 43/53. Its mission is to provide comprehensive scientific assessments of current scientific, technical and socio-economic information worldwide about the risk of climate change caused by human activity, its potential environmental and socio-economic consequences, and possible options for adapting to these consequences or mitigating the effects.

References

Albrecht A, Kandji ST (2003) Carbon sequestration in tropical agroforestry systems. Agr Ecosyst Environ 99(1-3):15–27

Atsumi S, Higashide W, Liao JC (2009) Direct photosynthetic recycling of carbon dioxide to isobutyraldehyde. Nat Biotechnol 27(12):1177–1180

Beatty PH, Good AG, Gates M (2011) Future prospects for cereals. Science 333(July):416–417

Borak B, Ort DR, Burbaum JJ (2013) Energy and carbon accounting to compare bioenergy crops. Curr Opin Biotechnol 24:1–7

Bouwman AF, Boumans LJM, Batjes NH (2001) Modeling global annual N_2O and NO emissions from fertilized fields. Glob Biogeochem Cy 16:1080
Buttel FH (2003) Editor's Choice Series on Agricultural Ethics Internalizing the Societal Costs of Agricultural Production. Plant Physiol 133(December):1656–1665
Cannell MGR (2003) Carbon sequestration and biomass energy offset : theoretical, potential and achievable capacities globally, in Europe and the UK. Biomass Bioenergy 24:97–116
Cannell MGR, Thornley JHM (1998) N-poor ecosystems may respond more to elevated [CO_2] than N-rich ones in the long term. A model analysis of grassland. Glob Chang Biol 4:431–442
Charpentier M, Oldroyd G (2010) How close are we to nitrogen-fixing cereals? Curr Opin Plant Biol 13(5):556–564
Chen X-P, Cui Z-L, Vitousek PM et al (2011) Integrated soil-crop system management for food security. Proc Natl Acad Sci U S A 108(16):6399–6404
Ciampitti IA, Vyn TJ (2011) A comprehensive study of plant density consequences on nitrogen uptake dynamics of maize plants from vegetative to reproductive stages. Field Crop Res 121(1):2–18
Clemens J, Ahlgrimm H (2001) Greenhouse gases from animal husbandry: mitigation options. Nutr Cycl Agroecosyst 68:287–300
Crews TE, Peoples MB (2005) Can the synchrony of nitrogen supply and crop demand be improved in legume and fertilizer-based agroecosystems? A review. Nutr Cycl Agroecosyst 72(2):101–120
Cui Z, Zhang F, Chen X et al (2008) On-farm evaluation of an in-season nitrogen management strategy based on soil Nmin test. Field Crop Res 105(1-2):48–55
Cui Z, Chen X, Zhang F (2010) Current nitrogen management status and measures to improve the intensive wheat–maize system in China. Ambio 39(5-6):376–384
Davidson EA (2009) The contribution of manure and fertilizer nitrogen to atmospheric nitrous oxide since 1860. Nat Geosci 2(9):659–662
Davidson EA (2012) Representative concentration pathways and mitigation scenarios for nitrous oxide. Environ Res Lett 7(2):1–7
Den Herder G, Van Isterdael G, Beeckman T, De Smet I (2010) The roots of a new green revolution. Trends Plant Sci 15(11):600–607
Drinkwater LE, Wagoner P, Sarrantonio M (1998) Legume-based cropping systems have reduced carbon and nitrogen losses. Nature 396:262–265
Duce RA, LaRoche J, Altieri K et al (2008) Impacts of atmospheric anthropogenic nitrogen on the open ocean. Science 320(5878):893–897
Duvick DN (2005a) Genetic progress in yield of United States Maize (Zea mays L.). Maydica 50:193–202
Duvick DN (2005b) The contribution of breeding (Zea Maysl.). Adv Agron 86:83–145
D Ehhalt, Prather M (2001) Atmospheric chemistry and greenhouse gases. In: Joos F, McFarland M (eds) Climate change 2001: the scientific basis, pp. 240–287
ENA (2011) The European Nitrogen assessment sources, effects and policy perspectives. Cambridge University Press, New York, NY
Fischer RA, Byerlee D, Edmeades GO (2009) Can technology deliver on the yield challenge to 2050 ? FAO expert meeting on how to feed the world in 2050 2009;(June):24–26
Franche C, Lindström K, Elmerich C (2008) Nitrogen-fixing bacteria associated with leguminous and non-leguminous plants. Plant and Soil 321(1-2):35–59
Frederiksen P, Maenpaa M (2007) Analysing and synthesising European legislation in relation to water: A Watersketch report under WP1. NERI, Watersketch
Geider RJ, Delucia EH, Falkowski PG et al (2001) Primary productivity of planet earth: biological determinants and physical constraints in terrestrial and aquatic habitats. Glob Chang Biol 7:849–882
Gherbi H, Markmann K, Svistoonoff S et al (2008) SymRK defines a common genetic basis for plant root endosymbioses with arbuscular mycorrhiza fungi, rhizobia, and Frankia bacteria. Proc Natl Acad Sci U S A 105(12):4928–4932

Giller KE, Witter E, Corbeels M, Tittonell P (2009) Conservation agriculture and smallholder farming in Africa: the heretics' view. Field Crop Res 114(1):23–34
Godfray HCJ, Beddington JR, Crute IR et al (2010) Food security: the challenge of feeding 9 billion people. Science (New York, NY) 327(5967):812–818
Good AG, Beatty PH (2011) Fertilizing nature: a tragedy of excess in the commons. PLoS Biol 9(8), e1001124
Good AG, Shrawat AK, Muench DG (2004) Can less yield more is reducing nutrient input into the environment compatible with maintaining crop production? Trends Plant Sci 9(12):597–605
Goulding K (2004) Pathways and losses of fertilizer nitrogen at different scales. In: Mosier AR, Syers KJ, Freney JR (eds) Agriculture and the nitrogen cycle: assessing the impacts of fertilizer use on food production and the environment. Island Press, Washington, DC, pp 209–219
Gruber N, Galloway JN (2008) An earth-system perspective of the global nitrogen cycle. Nature 451(7176):293–296
Hawkesford MJ (2012) Improving nutrient use efficiency in crops. John Wiley, Chichester
Ho S-H, Chen C-Y, Lee D-J, Chang J-S (2011) Perspectives on microalgal CO_2-emission mitigation systems – a review. Biotechnol Adv 29(2):189–198
IPCC (2001) Summary for policymakers: emissions scenarios, a special report of IPCC working group III. http://www.ipcc.ch/pdf/special-reports/spm/sres-en.pdf
IPCC (2005) IPCC special report on carbon dioxide capture and storage. Prepared by working group III of the Intergovernmental Panel on Climate Change. In: Metz B, Davidson O, de Coninck H, Loos M, Meyer L (eds) Carbon dioxide capture and storage. Cambridge University Press, Cambridge, p 442
IPCC (2007) Summary for policymakers. In: Solomon S, Qin D, Manning M et al (eds) Climate change 2007: the physical science basis Contribution of working group I to the fourth assessment report of the intergovernmental panel on climate change [Internet]. Cambridge University Press, Cambridge, pp 1–18
Kant S, Bi Y-M, Rothstein SJ (2011) Understanding plant response to nitrogen limitation for the improvement of crop nitrogen use efficiency. J Exp Bot 62(4):1499–1509
Kumar A, Ergas S, Yuan X et al (2010) Enhanced CO_2 fixation and biofuel production via microalgae: recent developments and future directions. Trends Biotechnol 28(7):371–380
Ladha JK, Reddy PM (2000) The quest for nitrogen fixation in rice. In: Ladha JK, Reddy PM (eds) Proceedings of the third working group meeting of the frontier project on assessing opportunities for nitrogen fixation in rice, 9–12 Aug 1999. International Rice Research Institute, Makati City, p 354
Lal R (2004) Soil carbon sequestration impacts on global climate change and food security. Science (New York, NY) 304(5677):1623–1627
Lal R, Follett RF, Kimble J, Cole CV (1999) Managing U.S. cropland to sequester carbon in soil. J Soil Water Conserv 54:374–381
Langridge P (2013) Investigate establishing an EWG on phenotyping – approach possible proposer. Interim Scientific Board lunch, San Diego, USA. http://www.wheatinitiative.org/sites/default/files/1301-scientific-board-minutes.pdf
Lawlor DW, Cornic G (2002) Photosynthetic carbon assimilation and associated metabolism in relation to water deficits in higher plants. Plant Cell Environ 44:275–294
Le Floch DM, Bereiter B, Blunier T et al (2013) Dome C 800,000-year record: European Project for Ice Coring in Antarctica (EPICA) members. http://cdiac.ornl.gov/trends/co2/ice_core_co2.html. Accessed 10 Apr 2013
Lee J-S, Lee J-P (2003) Review of advances in biological CO_2 mitigation technology. BIotechnol Bioproc Eng 8:354–359
Liepman AH, Olsen LJ (2003) Alanine aminotransferase homologs catalyze the glutamate: glyoxylate aminotransferase reaction in peroxisomes of Arabidopsis 1. Plant Physiol 131(January): 215–227
Long SP, Ort DR (2010) More than taking the heat: crops and global change. Curr Opin Plant Biol 13(3):241–248

Lucas PL, Van Vuuren DP, Olivier JGJ, Den Elzen MGJ (2007) Long-term reduction potential of non-CO_2 greenhouse gases. Environ Sci Policy 10(2):85–103
MacFarling Meure C, Etheridge D, Trudinger C et al (2006) Law Dome CO_2, CH_4 and N_2O ice core records extended to 2000 years BP. Geophys Res Lett 33(14), L14810
Mahdi SS, Hassan GI, Samoon SA, Rather HA, Dar SA (2010) Zehra6 B. Bio-fertilizers in organic agriculture. J Phytol 2(10):42–54
Maillet F, Poinsot V, André O et al (2011) Fungal lipochitooligosaccharide symbiotic signals in arbuscular mycorrhiza. Nature 469(7328):58–63
Masclaux-Daubresse C, Daniel-Vedele F, Dechorgnat J, Chardon F, Gaufichon L, Suzuki A (2010) Nitrogen uptake, assimilation and remobilization in plants: challenges for sustainable and productive agriculture. Ann Bot 105(7):1141–1157
McAllister CH, Beatty PH, Good AG (2012) Engineering nitrogen use efficient crop plants: the current status. Plant Biotechnol J 10(9):1011–1025
Merrick M, Dixon R (1984) Why don't plants fix nitrogen? Trends Biotechnol 2(6):162–166
Miller AJ, Fan X, Shen Q, Smith SJ (2007) Amino acids and nitrate as signals for the regulation of nitrogen acquisition. J Exp Bot 59(1):111–119
Miranowski J, Rosburg A (2011) US maize yield growth implications for ethanol and greenhouse gas emissions. AgBioForum 14(3):120–132
Montzka SA, Dlugokencky EJ, Butler JH (2013) Non-CO_2 greenhouse gases and climate change. Nature 476(7358):43–50
Näsholm T, Kielland K, Ganeteg U (2009) Uptake of organic nitrogen by plants. New Phytol 182(1):31–48
Njenga M, Gurung J (2011) Enhancing gender responsiveness in putting nitrogen to work for smallholder farmers in Africa (N2Africa). Women Organising for Change in Agriculture and NRM. pp 1–33. http://www.n2africa.org/sites/n2africa.org/files/images/images/WOCANGenderreport.pdf
Okumoto S, Pilot G (2011) Amino acid export in plants: a missing link in nitrogen cycling. Mol Plant 4(3):453–463
Oldroyd GED, Harrison MJ, Paszkowski U (2009) Reprogramming plant cells for endosymbiosis. Science (New York, NY) 324(5928):753–754
Parikh MR, Greene DN, Woods KK, Matsumura I (2006) Directed evolution of RuBisCO hypermorphs through genetic selection in engineered E. coli. Protein Eng Des Sel 19(3):113–119
Park S, Croteau P, Boering KA et al (2012) Trends and seasonal cycles in the isotopic composition of nitrous oxide since 1940. Nat Geosci 5(4):261–265
Parkin IAP, Clarke WE, Sidebottom C et al (2010) Towards unambiguous transcript mapping in the allotetraploid Brassica napus. Genome 53(11):929–938
Paustian K, Cole V (1998) CO_2 mitigation by agriculture: an overview. Clim Change 40: 135–162
Peoples MB, Boyer EW, Goulding KWT et al (2004) Pathways of nitrogen loss and their impacts on human health and the environment. In: Mosier AR, Syers JK, Freney JR (eds) Agriculture and the nitrogen cycle. The Scientific Committee on problems of the environment. Island Press, Covelo, CA, pp 53–69
Peoples MB, Brockwell J, Herridge DF et al (2009) The contributions of nitrogen-fixing crop legumes to the productivity of agricultural systems. Symbiosis 48(1-3):1–17
Perez I, Wire C (2013) Climate change and rising food prices heightened Arab Spring. Scientific American. http://www.scientificamerican.com/article.cfm?id=climate-change-and-rising-food-prices-heightened-arab-spring
Piao S, Ciais P, Huang Y et al (2010) The impacts of climate change on water resources and agriculture in China. Nature 467(7311):43–51
Reddy ASN, Day IS, Narasimhulu SB et al (2002) Isolation and characterization of a novel calmodulin-binding protein from potato. J Biol Chem 277(6):4206–4214
Rentsch D, Schmidt S, Tegeder M (2007) Transporters for uptake and allocation of organic nitrogen compounds in plants. FEBS Lett 581(12):2281–2289

Saikia SP, Jain V (2007) Biological nitrogen fixation with non-legumes: an achievable target or a dogma? Curr Sci 92(3):317–322

Schneider UA, McCarl BA (2003) Economic potential of biomass based fuels for greenhouse gas emission mitigation. Environ Resour Econ 24:291–312

Shanahan JF, Kitchen NR, Raun WR, Schepers JS (2008) Responsive in-season nitrogen management for cereals. Comput Electron Agr 61(1):51–62

Sitch S, Brovkin V, Von Bloh W, Van Vuuren D, Eickhout B, Ganopolski A (2005) Impacts of future land cover changes on atmospheric CO_2 and climate. Global Biogeochem Cycles 19(2):1–15

Smith KA, Conen F (2004) Impacts of land management on fluxes of trace greenhouse gases. Soil Use Manage 20(2):255–263

Stewart C, Hessami M-A (2005) A study of methods of carbon dioxide capture and sequestration – the sustainability of a photosynthetic bioreactor approach. Energ Convers Manage 46(3): 403–420

Stoltzfus JR, So R, Malarvithi PP, Ladha JK, Bruijn FJ (1997) Isolation of endophytic bacteria from rice and assessment of their potential for supplying rice with biologically fixed nitrogen. Plant and Soil 194:25–36

Tegeder M, Rentsch D (2010) Uptake and partitioning of amino acids and peptides. Mol Plant 3(6):997–1011

The EU Nitrates Directive. European Commission (2010) http://ec.europa.eu/environment/water/water–nitra

Thomas QR, Canham CD, Weathers KC, Goodale CL (2009) Increased tree carbon storage in response to nitrogen deposition in the US. Nat Geosci 3(1):13–17

UNFCCC (2003) Inventory of U.S. greenhouse gas emissions and sinks: 1990–2001. http://unfccc.int/national_reports/annex_i_ghg_inventories/national_inventories_submissions/items/618.php

UNFCCC (2013) Annual European Union greenhouse gas inventory 1990–2011 and inventory report 2013; submission to the UNFCCC Secretariat. http://unfccc.int/national_reports/annex_i_ghg_inventories/national_inventories_submissions/items/7383.php. Accessed 23 Apr 2013

UNFCCC (2013) Second national communication on climate change of The People's Republic of China. http://unfccc.int/essential_background/library/items/3599.php?rec=j&priref=7666#beg

US-EPA (2013) Inventory of U.S. greenhouse gas emissions and sinks: 1990–2011. http://www.epa.gov/climatechange/emissions/usinventoryreport.html

US-EPA. United States, Environmental Protection Agency (2013) http://www.epa.gov/glimaechange/ghgemissions/. Accessed 16 Apr 2013

USGRP (2009) Global climate change impacts in the United States [Internet]. Cambridge University Press, New York, NY

Van Groenigen K-J, Six J, Hungate BA, De Graaff M-A, Van Breemen N, Van Kessel C (2006) Element interactions limit soil carbon storage. Proc Natl Acad Sci U S A 103(17):6571–6574

WRI CAIT (2012) World Resources Institute, Climate Analysis Indicators Tool (WRI, CAIT). WRI CAIT, Washington, DC

Yanni YG, Dazzo FB (2010) Enhancement of rice production using endophytic strains of Rhizobium leguminosarum bv. trifolii in extensive field inoculation trials within the Egypt Nile delta. Plant and Soil 336(1-2):129–142

Yu C-H, Huang C-H, Tan C-S (2012) A review of CO_2 capture by absorption and adsorption. Aerosol Air Qual Res 12:745–769

The Impact of Genomics Technology on Adapting Plants to Climate Change

David Edwards

Introduction

Agriculture is a relatively recent development in the history of human evolution; however its impact on our species has been profound, enabling the establishment of increasingly large communities, supporting industrialisation and advances which have arguably led to humans becoming the dominant species on the planet. To date, advances in agriculture have generally matched increasing population growth, however as we are faced with unprecedented population size, combined with global climate change, there is an increasing risk that agriculture may not keep pace with human food requirements.

Increased global food production in recent decades has predominantly been achieved through the application of fertilisers, weed and pest control, improved agronomic practices and the breeding of improved varieties (Duvick 2005; Abberton et al. 2015). Continued increases using these methods are unsustainable, particularly with the reliance of nitrogen fertiliser on oil reserves and the finite resources of phosphorus fertiliser. The improvement of varieties is considered to offer the greatest potential for sustainable increases in food production, and substantial investment is being made in technologies which support and accelerate the development of advanced varieties.

Current global agriculture supports food production in a wide range of agro ecological zones, with a diversity of crop species which have been domesticated and bred to variable extents. However a limited number of key crops dominate the calorific component of the human diet. Some of these, such as maize and rice, have benefitted from advanced breeding technologies, however others, such as wheat, have lagged behind due to a combination of high genome complexity and limited

D. Edwards, Ph.D. (✉)
School of Plant Biology and Institute of Agriculture, University of Western Australia,
35 Stirling Highway, Crawley, WA 6009, Australia
e-mail: Dave.Edwards@uwa.edu.au

D. Edwards, J. Batley (eds.), *Plant Genomics and Climate Change*,
DOI 10.1007/978-1-4939-3536-9_8

provision for companies to secure a return on the large investment required for advanced genomics based crop improvement programs. There are yet many more crops which could benefit substantially from genomics based improvement, though there is little economic support for such investment, while the cost of genomics technology remains relatively high.

Genome Sequencing

Biological research has been revolutionised by the rapid advances in DNA sequencing that have occurred over the last 5–10 years (Shendure and Aiden 2012). While the traditional Sanger based method, first developed in the 1970s and later highly automated, was the technology of choice for the first Human and plant genome sequencing projects, it has now been mostly superseded by the advances in next generation DNA sequencing (NGS). The first commercial NGS system was demonstrated by 454 technologies and commercialised by Roche as the GS20, capable of sequencing over 20 million base pairs in just over 4 h. This was replaced in 2007 by the GSFLX model, capable of producing over 100 million base pairs of sequence in a similar amount of time. The Solexa sequencing system, which evolved into the Illumina Genome Analyser and later the Illumina HiSeq/MiSeq/NextSeq range have since come to dominate the sequencing market, with the disadvantage of relatively short reads being more than compensated by the large volume, low cost and high read accuracy compared to competing sequencing systems. The DNA sequencing system market is rapidly changing and highly competitive, with several companies such as Pac Bio (Pacific Biosciences) and Life Technologies gaining a share, and the prospect of further technologies, such as those developed by Oxford Nanopore, offering the potential to further revolutionise the generation of DNA sequence data. The DNA sequencing market also continues to grow, driven by the move in biomedical research towards highly profitable clinical applications. While agricultural applications remain relatively small in comparison, agricultural research and crop improvement still benefit greatly from these advances in technology.

The advances in DNA sequencing are impacting crop improvement in several ways. The production of reference genome assemblies provides insight into gene content, the ability to associate traits with specific genes, with subsequent understanding of the related biochemical and biophysical role of the genes in the expression of the trait. The re sequencing of diverse crop varieties provides evidence of gene content variation and DNA sequence differences between allelic variants, while the sequencing of expressed gene products provides information on where and when genes are functioning. This information, when integrated within breeding and selection approaches, offers the potential to accelerate the production of advanced climate resilient crops.

As increasing numbers of crop genomes are sequenced using higher throughput technologies, there has been a general decline in assembly quality. It is usual for draft genome assemblies to ignore highly repetitive portions of the genome, however some published assemblies also have major errors on the chromosomal scale which has the potential to mislead further analysis (Jain et al. 2013; Ruperao et al. 2014). Substantial investment is required to develop genome assembly validation and improvement methods to ensure that genome assemblies are fit for purpose. While many genome projects aim to produce finished 'gold standard' assemblies, these are not required for many applications such as the characterisation of gene content or the discovery of molecular genetic markers. As technology advances, an increased number of crops will benefit from finished gold standard assemblies, however the lack of resources to produce such assemblies for minor crops and relates species should not prevent the development of more limited genomic resources which can be applied to understand both crop evolution and for accelerated crop improvement.

Molecular Genetic Markers

The most commonly applied molecular tools for crop improvement are molecular genetic markers. The application of molecular markers to advance plant breeding is now well established. Modern agricultural breeding is dependent on molecular markers for the rapid and precise analysis of germplasm, trait mapping and marker assisted selection. Molecular markers can be used to select parental genotypes in breeding programs, eliminate linkage drag during back-crossing, and select for traits that are difficult to measure using phenotypic assays. Early molecular markers such as restriction fragment length polymorphisms (RFLPs) were superseded by PCR based assays including simple sequence repeats and amplified fragment length polymorphisms. Since the advent of NGS sequencing, single nucleotide polymorphisms (SNPs) have become the principal markers utilised in plant genetic analysis (Edwards et al. 2013).

SNPs are the ultimate form of molecular genetic marker, as a nucleotide base is the smallest unit of inheritance and a SNP represents a single nucleotide difference between two individuals. SNPs are direct markers as the sequence information provides the exact nature of the allelic variants. They also represent the most frequent type of genetic polymorphism and may therefore provide a high density of markers near a locus of interest (Edwards et al. 2007a). Many early SNP discover methods were laboratory based, however the expansion in the production of DNA sequence data has moved SNP discovery from the lab to the computer (Edwards et al. 2007b). The growth in genome sequences has also helped accelerate SNP discovery and linked these markers to the physical genome (Edwards et al. 2013; Visendi et al. 2013). This has changed the resolution of trait mapping from traditional genetic maps down to individual annotated genes on a physical map.

In addition to the revolution in SNP discovery, there has been a parallel revolution in the genotyping of these SNPs in populations. A range of SNP assays now exist which vary in throughput, and often focus on either the genotyping of a relatively small number of SNPs across a large population, or a large number of SNPs across a smaller population . Such high throughput technology includes Kaspar arrays and

Illumina infinium assays. More recently, and associated with the expansion of NGS technology, genotyping by sequencing (GBS) approaches have been developed (Elshire et al. 2011; Poland et al. 2012a). These have advantages in that they may concurrently discover and genotype SNPs in populations and provide dense genome wide genotype data. Early GBS approaches used reduced representation methods to reduce costs, however as sequencing costs continue to plummet, whole genome skim sequencing is becoming feasible, providing the highest resolution SNP genotyping in populations.

The Application of Genomics for New Breeding Methods

The rapid advances in genomics are now making an impact in the field of crop breeding. The use of genetic information through trait associated molecular markers has been gradually increasing as the technology has developed, however the recent explosion in genomic data is precipitating a fundamental shift towards genomics based breeding (Hamblin et al. 2011). Early forms of molecular markers were used for the construction of genetic linkage maps and the association of segregating markers with important agronomic traits. These markers could then be applied in selection during breeding, removing the requirement for often expensive and unreliable phenotypic selection. The ability to identify and genotype very large numbers of SNPs at ever reducing costs has enabled the expansion of marker assisted selection in breeding to a broader range of traits and across a wider range of crops (Varshney et al. 2014).

One of the impacts of the availability of high throughput genome wide markers is a move towards population based trait association and breeding. The move from biparental populations towards nested association mapping (NAM) (Yu et al. 2008; McMullen et al. 2009) or multiparent advanced generation intercross (MAGIC) (Rebetzke et al. 2014) based populations greatly increases the number of recombinations in the population with a subsequent increase in the resolution of trait mapping, often down to single or a few candidate genes. When combined with genome wide re-sequencing and mapping reads to reference genomes, it is possible to accurately delimit recombination positions and resolve gene and SNP level impacts on major traits. Once candidate genes are identified and validated, we gain a greater understanding of the biological mechanism underlying the trait and can approach trait improvement through marker based breeding or genetic modification. Furthermore, a precise understanding of the molecular basis of traits enables the engineering of novel alleles or the mining of potentially favourable alleles from wild germplasm or breeding diversity sets, supporting further enhancement of the trait.

The availability of large numbers of SNPs and the reducing cost of genome wide genotyping is leading to new approaches for breeding which utilise this whole genome diversity information. The genomic selection approach developed by Meuwissen et al. (2001) was first adopted in animal breeding and only recently been applied in crop breeding (Poland et al. 2012b; Resende et al. 2012). This approach offers significant potential to accelerate crop improvement and is complementary to the more traditional marker assisted breeding and candidate gene selection methods. Understanding the genomic diversity within crops and related species and linking this genomic diversity with potentially important agronomic traits is key to the development of improved crops with enhanced yield and resilience to predicted climate change scenarios.

Future Challenges

It is difficult to underestimate impact of the rapid advances that continue to occur in the field of genomics. The most recent burst of developments have been driven by the revolution in DNA sequencing which continues today and is likely to continue for the coming decades. We are only now starting to see the impact of genomics in the fields of personalised medicine and crop improvement. Draft genome sequences for the major crops have only recently been produced and many projects are underway around the world to define the diversity of gene content for these crops, producing crop reference pan-genomes. An increasing number of minor crop species and wild relatives are also being sequenced, providing insights into domestication and breeding bottlenecks. Relating these data to climate related agronomic traits for use in breeding remains a huge challenge, and one which will require coordination of diverse skills and expertise across crops, genomics technologies and bioinformatics, together with the substantial funding support required to accelerate translation of these genomics advances to secure future food supplies for the growing population in an unpredictable climate.

References

Abberton M, Batley J, Bentley A, Bryant J, Cai H, Cockram J, Costa de Oliveira A, Cseke L, Dempewolf H, De Pace C, Edwards D, Gepts P, Greenland A, Hall A, Henry R, Hori K, Howe G, Hughes S, Humphreys M, Lightfoot D, Marshall A, Mayes S, Nguyen H, Ogbonnaya F, Ortiz R, Paterson A, Tuberosa R, Valliyodan B, Varshney R, Yano M (2015) Global agricultural intensification during climate change: a role for genomics. Plant Biotechnol J. doi:10.1111/pbi.12467. [Epub ahead of print]

Duvick DN (2005) The contribution of breeding to yield advances in maize (*Zea mays* L.). Adv Agron 86:83–145

Edwards D, Forster JW, Chagné D, Batley J (2007a) What are SNPs? In: Oraguzie NC, Rikkerink EHA, Gardiner SE, De Silva HN (eds) Association mapping in plants. Springer, New York, NY, pp 41–52

Edwards D, Batley J, Cogan NOI, Forster JW, Chagné D (2007b) Single nucleotide polymorphism discovery. In: Oraguzie NC, Rikkerink EHA, Gardiner SE, De Silva HN (eds) Association mapping in plants. Springer, New York, NY, pp 53–76

Edwards D, Batley J, Snowdon R (2013) Accessing complex crop genomes with next-generation sequencing. Theor Appl Genet 126:1–11

Elshire RJ, Glaubitz JC, Sun Q, Poland JA, Kawamoto K, Buckler ES, Mitchell SE (2011) A robust, simple genotyping-by-sequencing (GBS) approach for high diversity species. PLoS One 6, e19379

Hamblin MT, Buckler ES, Jannick JL (2011) Population genetics of genomics-based crop improvement methods. Trends Genet 27:98–106

Jain M, Misra G, Patel RK, Priya P, Jhanwar S, Khan AW, Shah N, Singh VK, Garg R, Jeena G, Yadav M, Kant C, Sharma P, Yadav G, Bhatia S, Tyagi AK, Chattopadhyay D (2013) A draft genome sequence of the pulse crop chickpea (*Cicer arietinum* L.). Plant J 74:715–729

McMullen MD, Kresovich S, Villeda HS, Bradbury P, Li H, Sun Q, Flint-Garcia S, Thornsberry J, Acharya C, Bottoms C, Brown P, Browne C, Eller M, Guill K, Harjes C, Kroon D, Lepak N, Mitchell SE, Peterson B, Pressoir G, Romero S, Oropeza Rosas M, Salvo S, Yates H, Hanson M, Jones E, Smith S, Glaubitz JC, Goodman M, Ware D, Holland JB, Buckler ES (2009) Genetic properties of the maize nested association mapping population. Science 325:737–740

Meuwissen THE, Hayes BJ, Goddard ME (2001) Prediction of total genetic value using genome-wide dense marker maps. Genetics 157:1819–1829

Poland JA, Brown PJ, Sorrells ME, Jannick J-L (2012a) Development of high-density genetic maps for barley and wheat using a novel two-enzyme genotyping-by-sequencing approach. PLoS One 7, e32253

Poland JA, Endelman J, Dawson J, Rutkowski J, Wu S, Manes Y, Driesigacker S, Crossa J, Sanchez-Villeda H, Sorrells M, Jannick J-L (2012b) Genomic selection in wheat breeding using genotyping-by-sequencing. Plant Gene 5:103–113

Rebetzke GJ, Verbyla AP, Verbyla KL, Morell MK, Cavanagh CR (2014) Use of a large multiparent wheat mapping population in genomic dissection of coleoptile and seedling growth. Plant Biotechnol J 12:219–230

Resende MDV, Resende MF Jr, Sansaloni CP, Petroli CD, Missiaggia AA, Aguiar AM, Abad JM, Takahashi EK, Rosado AM, Faria DA, Pappas GJ Jr, Kilian A, Grattapaglia D (2012) Genomic selection for growth and wood quality in eucalyptus: capturing the missing heritability and accelerating breeding for complex traits in forest trees. New Phytol 194:116–128

Ruperao P, Chan KCK, Azam S, Karafiátová M, Hayashi S, Čížková J, Saxena RK, Šimková H, Song C, Vrána J, Chitikineni A, Visendi P, Gaur PM, Millán T, Singh KB, Taran B, Wang J, Batley J, Doležel J, Varshney RK, Edwards D (2014) A chromosomal genomics approach to assess and validate the Desi and Kabuli draft chickpea genome assemblies. Plant Biotechnol J 12:778–786

Shendure J, Aiden EL (2012) The expanding scope of DNA sequencing. Nat Biotechnol 30:1084–1094

Varshney RK, Terauchi R, McCouch SR (2014) Harvesting the promising fruits of genomics: applying genome sequencing technologies to crop breeding. PLoS Biol 12, e1001883

Visendi PJ, Batley J, Edwards D (2013) Next generation characterisation of cereal genomes for marker discovery. Biology 2:1357–1377

Yu J, Holland JB, McMullen MD, Buckler ES (2008) Genetic design and statistical power of nested association mapping in maize. Genetics 178:539–551

Genomics of Salinity

Philipp Emanuel Bayer

Introduction

Salts (mostly sodium and chloride) occur naturally in all soils at various levels. Salinity is one of the major abiotic stresses faced by plants, together with drought and extreme temperatures. Salinity stress leads to loss of water, which in turn results in a loss of yield. Worldwide, more than 800 million ha of land are affected by salinity (FAO 2005), and in the next 25 years, salinity is expected to cause the loss of 30 % arable land, which will increase to 50 % by 2050 (Wang et al. 2003). Worldwide, Australia is one of the countries worst-affected by salinity. The clearing of native species by European farmers has led to less rainfall being used by plants (George et al. 2012). This in turn led to a rise of groundwater levels, which prior to the arrival of Europeans had been stable, and salt dissolved in the groundwater is being carried to the surface, leading to an increase in soil salinity. Dry land salinity costs Australia roughly $3.5 billion per year, with an income loss of $200 million per year for farmers (Warrick 2006).

Plants have developed a variety of techniques to cope with salinity stress and salinity tolerance varies between crops and varieties. In plants, reaction to drought and salinity is usually measured using a few standard metrics (see for example Tambussi et al. 2007). The net CO_2 assimilated by photosynthesis is denoted by A. Water loss, for example due to transpiration or osmotic pressure, is denoted by T. Both A and T refer to the same fixed period. The quotient of both, A/T, is called the water use efficiency (commonly abbreviated as *WUE*). The higher the *WUE*, the more CO_2 is assimilated, or less water is lost. Halophytes tend to have a higher *WUE* than non-halophytes. For different studies, there are different definitions of *WUE* depending on the scale observed and the background of the researcher. Agronomists and crop

P.E. Bayer, B.Sc. (✉)
School of Agriculture and Plant Sciences, University of Queensland,
83 Harley Teakle Building, Brisbane, QLD 4072, Australia
e-mail: philipp.bayer@uwa.edu.au

D. Edwards, J. Batley (eds.), *Plant Genomics and Climate Change*,
DOI 10.1007/978-1-4939-3536-9_9

physiologists usually define *WUE* as the accumulated dry matter divided by water use in the same time period. Here, we will use the definition first presented.

Soil salinity is measured by the NaCl concentration (mM, millimolar). Officially, water salinity is measured using only one metric, however several metrics are used in the literature. The two most common are: Practical Salinity (Practical Salinity Scale 1978 (PSS-78)), and the official measurement; Absolute Salinity (Thermodynamic Equation of Seawater 2010 (TEOS-10)) (Pawlowicz 2010). The official symbol for Practical Salinity is S_P and the official symbol for Absolute Salinity is S_A. S_P has no unit, but is still often denoted as *psu*, which is officially discouraged (Pawlowicz 2010). The unit of S_A is g kg^{-1}. S_P is determined by measuring the electrical conductivity of a solution. The more salt is included in the solution, the higher the electrical conductivity is, resulting in a higher S_P. S_A is calculated by combining S_P with information to account for regional changes in seawater composition. Most studies cited in this chapter use S_P—if measurements were reported as *psu*, we repeat them here as S_P. As a reference point, seawater from the deep North Pacific has a S_P of roughly 34.836 and an S_A of 35.02 g kg^{-1} (Pawlowicz 2010). We will come back to these values in later sections.

Worldwide, average temperatures are on the rise due to an anthropogenic effect known as global warming. There are varying scenarios on exactly how much temperatures are expected to rise, which depends on how well humanity is able to tackle climate change (Meehl and Stocker 2007). In the lowest emissions scenario, global temperatures are expected to rise 1.1–2.9 °C, and 2.4–6.4 °C in the highest emission scenario. Both scenarios see an increase in water evaporation which leads to an increase in soil salinity.

With the rise of temperatures due to global warming, it is expected that there will be increased irrigation of farmland. This leads to increases in soil salinity (Utset and Borroto 2001). This chapter will summarize the current state of research into the mechanisms of salinity tolerance by several important crop species. Research into the genetic background of salinity tolerance is summarized, and current efforts to produce salinity tolerant crop cultivars are discussed.

Salinity Tolerance Mechanisms

Salinity stress can be divided into two subgroups: osmotic stress (also called “osmotic response”) and ionic stress (also called “salt-specific response”), both of which lead to secondary stresses like oxidation (Zhu 2001). Osmotic stress is caused by the inability of roots to take up water since the salt concentration outside of the root is higher than on the inside. Ionic stress occurs when high salinity concentrations inside the plant become toxic. Cells can compartmentalize salt in their vacuoles, but the capacity has an upper limit. Once this limit is reached, ions build up in the cytoplasm. Since the cytoplasm has a much smaller volume than the vacuole, the ratio of salt build-up is much higher. Over-abundance of ions in the cytoplasm leads to a variety of problems, including the inhibition of protein activity.

Both osmotic and ionic stress appear together, but in succession. The reaction to osmotic stress starts immediately in all species, regardless of whether they are classified as halophytes or not (Munns 2002). Ionic stress is delayed since the plant first has to accumulate toxic levels of salt in the cells. This takes a few days in salinity susceptible species, like lupin, and several weeks for more salt tolerant species like wheat and barley (Munns et al. 1995). The onset of ionic stress occurs later in slow-growing species than in fast-growing species. Salinity tolerance is a physiologically and genetically complex trait (Flowers 2004). There are three main mechanisms for plants to react to increased salinity. These are: increased tolerance to osmotic stress, active ion (Na^+ and Cl^-) exclusion, and increased tolerance to accumulated Na^+ or Cl^-. This section will explain their general function, for a detailed description, see Munns and Tester (2008).

Osmotic stress is caused by a higher osmotic pressure outside roots, leading roots to lose water. Plants have found several ways to deal with osmotic stress. First of all, osmotic stress leads to stomata closing and less cell expansion in root tips and young leaves. Both of these reactions are to the plant's benefit in some situations, and detrimental in others. For example, closed stomata in low water conditions lead to less water-loss; but in high water conditions, gas exchange may be disturbed. Similarly, less cell expansion means reduced yield. Plants also increase the synthesis of osmotically active solutes like proline to adjust the rate of osmosis in order to minimize loss of water (Jaarsma et al. 2013).

Ion exclusion is the active removal of Na^+ and Cl^- by transport proteins, especially from roots. Salt inhibits the activity of proteins (see for example Giberti et al. 2014). Na^+ competes with K^+ in binding to more than 50 proteins, but Na^+ cannot activate these proteins (Bhandal and Mahlik 1988). Salt tolerant tissue uses proteins like Na^+/H^+ antiporters to actively exclude salt. Plants also downregulate the long distance transport of salts, and transport less salts into plastids.

An elevated tolerance of tissue to elevated ion levels is another mechanism with which plants defend themselves. High ion concentrations in leaves have toxic effects. Older leaves die, since older leaves have more time to accumulate salts. Many different genes and gene families are involved in ion exclusion. One example are *HKT* (High-affinity K^+ transporter) genes, which are a family of genes that are linked to the removal of Na^+ from the xylem. Later sections include specific examples of the function of *HKT* in different plants.

How these three mechanisms are used, and how much resource a species uses for each mechanism varies considerably in between species. The following gives an overview of what we know about different mechanisms and genes involved in salinity tolerance in different crop species as well as model species.

Recent Examples in Crop Species

Low genetic variation within many domesticated crop species increases the difficulty of conventional breeding (Ashraf and Akram 2009). Traditionally, plant breeders focused on improving yield instead of improving tolerance to various stresses.

Salinity tolerance is a costly trait, and so breeding for increased yield may have selected against salinity tolerance (Roy et al. 2011). Since genetic variation in current crop varieties is so low, research has focused on introducing genes for increased salt tolerance from wild salt-tolerant relatives into their domesticated counterparts.

Arabidopsis thaliana

Due to its small genome, fast maturation time, small size, and ease of handling *Arabidopsis thaliana* has become the model species of plant biology. However, *A. thaliana* is one of the least salinity tolerant plants (Munns and Tester 2008) and the least salinity tolerant plant presented in this chapter.

In *A. thaliana*, several genes linked to salinity tolerance have been identified. One study screened 675,500 mutant seeds, out of which 16 lines could germinate in high salinity conditions (Quesada et al. 2000). Comparing gene expression in these 16 lines, four salinity tolerance genes could be identified. These were named numerically from *SAN1* to *SAN5*. Mutants with the genetic background from the Columbia line had less problems growing under saline conditions than mutants from the Landsberg line, indicating that the Columbia line carries different salinity tolerance genes or alleles.

In *Arabidopsis*, the *AtNHX* gene family is involved in Na^+ ion compartmentalization and maintaining intracellular K^+ status (Yokoi et al. 2002). There have been experiments to introduce the *Arabidopsis* Na^+/H^+ antiporter *AtNHX1* into crop species such as tomato (Zhang and Blumwald 2001). In tomato, overexpression of *AtNHX1* led to increased salinity tolerance compared to the wild type. Old and young leaves of transgenic plants exhibited a 20- to 28-fold increase in Na^+ content during salt stress. It is important to note that salt did not accumulate in fruits. The K^+ content decreased markedly in all leaves. This shows that in some plants and using some genes, overexpression of just a single gene can lead to remarkable salinity tolerance.

Another study monitored the expression of 7000 genes using a cDNA microarray under differing stress conditions, including salinity (Seki et al. 2002). The transcript abundance of 194 genes increased during high salinity conditions. Of these, the majority were transcription factors, genes coding for transport proteins and genes involved in metabolism. Down-regulated genes were mostly involved in photosynthesis, reflecting the stress impact of salinity treatment. Most of the genes involved in stress tolerance were exclusive to their specific stress, which the authors interpreted as meaning that the stress reaction genetic networks overlap only a little.

In *Arabidopsis*, the *HKT* gene family plays a role in the active removal of Na^+ from the xylem parenchyma and is linked to improved salt tolerance (Sunarpi et al. 2005). During salt stress, the expression of *AtHKT1* rises. Mutants carrying defective *AtHKT1* alleles carry substantially more Na^+ in their xylem and less in the phloem. This means that the expression of *AtHKT* ensures that less Na^+ is delivered to leaves during salt stress, thereby protecting leaves from salt toxicity.

Several other salt stress related genes have been identified in *A. thaliana* with three genes being particular important: *SOS1*, *SOS2* and *SOS3*. *SOS1* (for "salt overly sensitive1") encodes a Na^+/H^+ antiporter which is involved in the compartmentalization of salt into the vacuole (Shi et al. 2002). Plants overexpressing *SOS1* survive salinity stress longer, since these plants can safely accumulate Na^+ in the vacuole, thereby avoiding salt toxicity in the rest of the cell. There is evidence that *SOS1* works bidirectional: it can load salt into the xylem under low and medium salinity stress, and it can unload salts from the xylem under high salinity stress. Of the three best-characterized *SOS*-genes, *SOS1* may play the most important role, since mutants lacking *SOS1* are more susceptible to salinity stress than mutants lacking *SOS2* or *SOS3* (Zhu et al. 1998).

SOS2 encodes a serine/threonine type protein kinase which is involved in salinity tolerance and the presence of salt leads to the upregulation of *SOS2* in *A. thaliana* (Liu et al. 2000). Plants with mutated *SOS2* genes are not tolerant to high Na^+ stress or low K^+ stress. *SOS2* phosphorylates other proteins involved in salt stress tolerance—it is known that *SOS2* physically interacts with *SOS3* (Halfter et al. 2000), which means that both are involved in the same Na^+/K^+ homeostasis pathway. *SOS3* is a calcium sensing protein (Ishitani et al. 2000). It is most likely that *SOS3* is a Ca^{2+}-dependent protein that is involved in long-distance cell signaling. In plants lacking *SOS3* or *SOS2*, no upregulation of *SOS1* could be detected (Shi et al. 2000). These results suggest that *SOS1*, *SOS2* and *SOS3* are all involved in the same or similar pathways used to respond to salinity stress. One study generated several transgenic lines with *AtNHX1*, *SOS1*, *SOS2* and *SOS3* being overexpressed in all possible combinations, and tested these lines under salt stress conditions (Yang et al. 2009). Plants overexpressing *SOS3*, *SOS2*+*SOS3*, *AtNHX1*+*SOS3* and all three *SOS*-genes accumulated less Na^+. Transgenic plants overexpressing *AtNHX1* did not exhibit increased salinity tolerance compared to the wild type, which is contrary to previous results (for example Zhang and Blumwald (2001)). Plants overexpressing all three *SOS* genes did not fare better compared to plants overexpressing *SOS1* or *SOS3* alone.

More recent efforts have investigated the salinity tolerance of *A. thaliana* accessions from all over the world in order to assess genetic diversity (DeRose-Wilson and Gaut 2011). Strong diversity exists in the tested 96 *Arabidopsis* lines with some lines being unable to germinate even under moderate salinity conditions. A search for salinity tolerance-related QTL led to the identification of several candidate genes, some of which were known before this particular study. These candidate genes are involved in ABA biosynthesis, freezing tolerance, osmotic stress response, and temperature acclimation. This study adds to the body of evidence that salinity tolerance is a complex trait in which several genes and genetic pathways are involved. Some lines exhibit very low germination during salinity stress (around 30 %), but high survival for seeds that did germinate (around 90%) which suggests that different pathways are involved in salinity stress during the lifecycle of *A. thaliana*.

Even though *Arabidopsis* exhibits only little to moderate salt tolerance there have been efforts to introduce genes linked to salt tolerance in *Arabidopsis* into other plants. One example is the introduction of the salt tolerance *Arabidopsis thaliana*

HARDY transcription factor gene into *Trifolium alexandrium* (Egyptian clover) (Abogadallah et al. 2011). *T. alexandrium* is not tolerant to salt stress since it accumulates excessive amounts of Na^+ in the leaves leading to enzyme inhibition. Introduction and overexpression of *HARDY* into two transgenic *T. alexandrium* lines led to improved salt stress tolerance. Both lines showed significantly lower transpiration and stomatal conductance as well as lower Na^+ contents in their leaves compared to the wild type during salt and drought stress.

Both transgenic lines were tested in the field for comparison with the wild type. Under control conditions, both transgenic and wild type lines showed similar growth. Under salt stress, both transgenic lines had significantly higher fresh and dry weight. However, photosynthesis in the transgenic lines was similar between both lines at the end of the daily drought cycle. It seems that *HARDY* is involved in reduced uptake and transport of Na^+, thereby protecting photosynthetic enzymes from salt inhibition.

A more salt tolerant alternative to *A. thaliana* is the salt cress *Thellungiella halophila*. *T. halophila* is a close relative of *A. thaliana* which should be a better candidate as a source for salt tolerant genes (Inan et al. 2004). Salt cress grows rapidly under low salt stress and can still survive at near-sea water salt concentrations. Salt cress has a small genome, about twice the size of the *Arabidopsis* genome. One study successfully introduced salinity-related genes from *T. halophila* into Arabidopsis (Du et al. 2008). An entire cDNA library of salt cress was introduced into *Arabidopsis* to generate 125,000 lines, of which roughly 10 % exhibited salt tolerance. In the salt tolerant lines, several cDNAs were identified with little or no similarity to any known *Arabidopsis* genes. The predicted functions of these cDNAs were transcription factors, photosynthesis, chaperones and damage-protection. 30 % of the isolated cDNAs had no known function or known domains, indicating that salt cress carries unique and previously unknown salt tolerance related genes. Two novel genes called *ST225* and *ST6-66* were identified in highly tolerant transformed lines. *ST6-66* has a 79 % nucleotide similarity with the *Arabidopsis* homologue *At1g13930*, which is involved in salt stress tolerance, suggesting that *ST6-66* is an orthologue of *At1g13930*.

The assembly of the salt cress relative *Thellungiella salsuginea* genome revealed further salinity tolerance related genes (Wu et al. 2012). There are additional copies of genes encoding Na^+/K^+ transporters, genes for ABA biosynthesis, and other salt stress related gene families than there are copies in *Arabidopsis*, which explains the increased salinity tolerance. Further annotation and characterization of unknown genes as well as the assembly of the genomes of other salt cress relatives may reveal more possible sources of salinity tolerance.

The *T. salsuginea* genome carries more copies of *RAV* genes than *Arabidopsis*. This implies that *RAV* genes may be a source of salinity tolerance—rice and maize are also known to carry *RAV* genes. One study introduced the maize transcription factor *ZmRAV1* into transgenic *Arabidopsis* lines (Min et al. 2014). Under salt stress, transformed lines showed enhanced salt and osmotic stress tolerance, longer primary roots, increased fresh weight, and less cell membrane damage (measured by electrolyte loss). Further analysis showed that *ZmRAV1* is involved in upregulating

ROS (reactive oxygen stasis) homeostasis genes. It is known that salt stress leads to more ROS so it makes sense that upregulation of ROS homeostasis genes is helpful during salt stress.

*Wheat (***Triticum aestivum** *and* **T. turgidum***)*

Bread wheat (*Triticum aestivum*) is one of the most important crop species in the world. Accordingly, much research has been published on salinity tolerance in wheat. Both bread and durum wheat have a higher salinity tolerance than rice and *Arabidopsis*, but are less tolerant than the salt- tolerant relative *Thinopyrum ponticum* or barley (Munns and Tester 2008). In hexaploid bread wheat (AABBDD) salinity tolerance is located on the D genome, specifically on chromosome 4D (Gorham et al. 1987). Later studies used QTL analysis to confirm the location of the *Kna1* locus on 4DL (Dubcovsky et al. 1996), and bread wheat plants lacking a copy of *Kna1* exhibit lower salinity stress tolerance.

One study tested the salinity tolerance of 59 different bread wheat genotypes (Genc et al. 2007). Salinity tolerance and the salinity tolerance mechanisms, Na^+ exclusion and tissue tolerance, varied between genotypes. However, there was no direct link between increased Na^+ exclusion and tolerance, and no direct link between tissue tolerance and salinity tolerance. These results suggest that in wheat, as in other plants, salinity tolerance is a complex trait controlled by several different networks.

Durum wheat (*Triticum turgidum*) is less salt tolerant than bread wheat. Durum wheat is tetraploid (AABB) and does not have the D genome of bread wheat, so the genes responsible for salinity tolerance in bread wheat are absent in durum wheat. One study screened several durum wheat lines for salinity tolerance and found large genetic variance, to the point that some lines exhibited salinity tolerance close or identical to bread wheat (Munns et al. 1999). Tolerant lines exhibited low Na^+ concentration in leaves hinting towards the existence of effective Na^+ exclusion. The line with the highest tolerance was later shown to carry two loci on chromosome 2A linked to Na^+ exclusion (Munns et al. 2003). These loci were termed *Nax1* and *Nax2*. The presence of both loci decreases Na^+ transport from root to shoot by different mechanisms. *Nax1* is linked to removal of Na^+ in both root and leaf, while *Nax2* is only linked to less Na^+ uptake in the root. This means that *Nax2* may be homologous to *Kna1*, while *Nax1* may be a novel gene. Both *Nax1* and *Nax2* are present in the wild ancestor *Triticum monococcum* (James et al. 2006).

Nax2 has been successfully transferred into bread wheat (James et al. 2006). The gene underlying *Nax2* encodes an *HKT*-like transporter and was termed *TmHKT1;5-A*. Bread wheat lines carrying *TmHKT1;5-A* exhibited low Na^+ concentration in leaves compared to the wild type. Under control conditions, both wild type and transformed lines showed little or no difference in yield. However, under high salinity conditions, yield of the wild type decreased by 50 %, while yield of the transformed line decreased by only 36 %. This suggests that transgenic bread wheat plants carrying the *TmHKT1;5-A* gene are better adapted to high salinity conditions.

Another potential source of salinity tolerance for *T. aestivum* is *Thinopyrum ponticum* (tall wheatgrass), a salt tolerant relative of *Triticum*. There have been efforts to transfer the salinity tolerance from *Thinopyrum* to *Triticum*. Somatic hybridization was used to transfer tolerance traits to *Triticum*, leading to salt tolerant *Triticum/Thinopyrum* hybrid lines (Suiyun et al. 2004). The hybrid lines exhibited a salinity tolerance intermediate between the tolerance of *Triticum* and the tolerance of *Thinopyrum*. Further analysis revealed differences between the hybrid line (called Shanrong Nr. 3, or SR3) and the *Triticum* ancestor line Jinan 177 (Wang et al. 2008). Under control conditions, SR3 carried significantly more K^+ and Na^+ in roots than Jinan 177. Under salt stress conditions, SR3 exhibited significantly higher fresh weight, dry weight as well as significantly higher concentrations of Na^+ and K^+ in leaves and stems, but not in roots. This hints towards a better tolerance of SR3 to salt stress due to a higher tolerance to acquired Na^+/K^+ in leaves and stems. Proteomic analysis of SR3 revealed an abundance of differentially regulated proteins. Many different transcription factors were affected relating to signal transduction, transport, chaperones, carbon/nitrogen metabolism and detoxifying mechanisms. Mitochondrial and chloroplast proteins showed differences too, so non-nucleic genes were also transferred during the hybridization event that created SR3.

SR3 also carries the protease inhibitor *WRSI15* (Shan et al. 2008). Under increased salinity conditions, *WRSI15* is overexpressed in root tips of SR3, but not in the wild type. The wild type carried more Na^+ in the roots than SR3, which suggests that *WRSI15* is involved in Na^+ transport, possibly by protecting Na^+ transport related proteins from degradation by proteases.

Maize (Zea mays*)*

Like wheat, maize is one of the most important crop plants. Traditional breeding has led to some success in drought tolerance (reviewed in Campos et al. 2004). Several elite lines specifically bred for salt-tolerance are on the market. As with tomato, *Brassica napus* and *B. juncea AtNHX1* has been introduced into maize (Yin et al. 2004). Maize itself has low Na^+/H^+ antiporter activity, making it a good target for the introduction of *AtNHX1*. As expected, transgenic lines carrying *AtNHX1* exhibit higher salinity tolerance than non-transgenic lines. However, not all progeny of the transgenic lines exhibit elevated salinity tolerance, even though the gene was present. The authors explain the differences with transgene splicing, i.e., *AtNHX1* was somehow made inactive between the original lines and the offspring. It could also be that the natural genetic variance in offspring interacted with salinity tolerance traits, leading to less salinity tolerance regardless of the introduced *AtNHX1* gene. As outlined above, salinity tolerance is a multigenic trait influenced by many factors.

At least one study tried to introduce *AtNHX1* into maize without the need for marker genes in transgenic lines (Li et al. 2010). As expected, under control conditions wild type and transgenic lines behaved identically, but under salinity stress all transgenic lines outperformed the wild type in yield and the transgenic lines carried more Na^+ in leaves.

*Rice (***Oryza sativa***)*

Rice is a staple food in many parts of the world, and not surprisingly, much research has focused on salinity tolerance in rice. In rice, several salt tolerant mutants have been discovered. Monitoring of expression profiles of *Oryza sativa* during different stresses led to the discovery of 57 genes that were upregulated only in high salinity conditions (Rabbani et al. 2003). The expression of 15 genes was induced by all four observed stresses which suggest the existence of inter-connected stress-tolerance genetic pathways.

The genetic networks involved in *O. sativa* are similar to those of *A. thaliana* - of the 73 stress-inducible rice genes, 51 had already been reported in *A. thaliana*, even though both plants have evolved separately for 1 MYA. This may mean that these genetic pathways are highly conserved in between both species. There have been efforts to generate genetically modified *O. sativa* plants with increased salt tolerance. To this end, the Na^+/H^+ antiporter *AgNHX1* has been transferred from the halophyte *Atriplex gmelini* to rice (Ohta et al. 2002). Transgenic lines carrying the gene survived 3 days under high salinity stress conditions which killed the wild type. However, 3 days survival is generally not long enough for plants which have to grow and produce yield in saline conditions. Only younger leaves survived leaf salinity treatments, since older leaves already accumulated too much salt in the vacuole. This indicates that *AgNHX1* alone is not enough to establish increased salinity tolerance.

The Arabidopsis gene salt tolerance gene *HARDY* has been successfully introduced into rice (Karaba et al. 2007). Mutants carrying the *HARDY* gene exhibited a 50–100 % higher water use efficiency (*WUE*) during well watered control conditions, and a 50 % higher *WUE* under drought stress. As a side-effect, roots of mutants were harder to remove from the soil, and leaves were smaller and thicker than wild type leaves.

One study characterized 100 SSRs in 140 recombinant inbred lines (RIL) in order to find salinity tolerance QTL (Thomson et al. 2010). The *Saltol* QTL was identified on chromosome 1, which is linked to seedling stage salinity tolerance by influencing Na^+/K^+ homeostasis. The exact gene underlying the *Saltol* QTL is probably the sodium transporter *SKC1*, which unloads Na^+ from the xylem (Ren et al. 2005). *SKC1* is similar to *AtHKT1* and thus belongs to the *HKT* gene family.

Another source of genetic salinity tolerance is C_4 photosynthesis. The development of C_4 photosynthesis is linked to an increase of salinity tolerance (see section "Halophytes and Extremophiles"), and there are ongoing efforts to transfer C_4

photosynthesis into rice, a C_3 plant, in order to increase yield and increase the rate of fixated CO_2 (von Caemmerer et al. 2012). Successful introduction of C_4 photosynthesis into rice may lead to increased salinity tolerance as a side-effect. At the time of writing, no C_4 rice has been produced, but researchers expect C_4 rice plants by 2016.

Brassica

The genus *Brassica* contains many important crop species, some of which share genomes. They include the diploid *Brassica oleracea* (cabbage, broccoli, C-genome), *B. rapa* (Chinese cabbage, turnip, A-genome) and *B. nigra* (black mustard, B-genome). Combinations of these genomes lead to the amphidiploid *B. juncea* (Indian mustard, AB-genome), *B. napus* (canola, rapeseed, AC-genome) and *B. carinata* (Ethiopian mustard, BC-genome).

In general, all *Brassicas* are moderately salt tolerant, with the amphidiploids being slightly more tolerant than the diploids, and the widely cultivated *B. napus* exhibiting the highest salt tolerance, and *B. nigra* exhibiting the lowest salinity tolerance (He and Cramer 1992). *B. napus* is one of the most economically important *Brassica* species, so most salinity tolerance research focused on this species.

Different *B. napus* cultivars react differently to salinity stress (Tunçtürk et al. 2013). A salt tolerance gene specific to *B. napus* called *BnD22* is induced by salinity and water stress (Reviron et al. 1992). *BnD22* is a protease inhibitor which delays the onset of leaf senescence by protecting proteins from premature degradation (Ilami et al. 1997). The presence of *BnD22* may explain why *B. napus* is more salt tolerant than other *Brassicas*. There is no evidence that *BnD22* exists in other *Brassicas*, although it may exist in *B. rapa* or *B. oleracea* due to the shared A and C genomes.

The gene *AtNHX1* was successfully introduced from *A. thaliana* into *B. napus* (Zhang et al. 2001). The higher Na^+ content in leaves of transgenic lines suggests that *AtNHX1* is involved in storing excess Na^+ in the vacuole, thereby conferring increased salinity tolerance until the vacuole is full. Other efforts to generate transgenic *B. napus* lines overexpressing *AtNHX1* have shown that salt stress increased the uptake of sulphur which lead to an increased biosynthesis of cysteine and glutathione (Ruiz and Blumwald 2002). Both cysteine and glutathione are used to protect the plant from active oxygen species and other stresses.

Transgenic lines of *B. juncea* with higher salinity tolerance have been generated by introducing the *PgNHX1* gene from *Pennisetum glaucum* (Rajagopal et al. 2006), a Na^+/H^+ antiporter which is responsible for vacuolar compartmentalization of Na^+. Transgenic lines carrying the gene exhibited a higher salinity tolerance than the wild type. As expected, transgenic lines exhibit higher concentration of Na^+ in vacuoles, and therefore younger leaves exhibit a higher salinity tolerance than older leaves.

*Barley (*Hordeum vulgare*)*

Barley is a major cereal grain mostly used for brewing, but also for barley bread and in soups. Differentially expressed genes between salt stress and control plants have been identified (Ueda et al. 2002). Upregulated genes seem to be concentrated in roots, and most overexpressed genes are involved in signal transduction and stress tolerance. The transcription factor *SCARECROW*, which was originally found to be related to plant tissue morphogenesis, is overexpressed in barley under salt stress. A few *SCARECROW*-like genes (called *SLCs*) have since been identified as being involved in salinity tolerance, for example in *A. thaliana* (Ma et al. 2010). Recent data has shown that *SCARECROW* plays a major role in C_4 photosynthesis in maize (Slewinski et al. 2012). However, many upregulated genes are still of unknown function.

Another study used microarrays to measure gene expression in salt-stressed and drought-stressed barley (Oztur et al. 2002). Upregulated transcripts overlapped little between drought and salt stress, which suggests that in barley, genetic networks involved in both stresses don't overlap significantly. An enzyme involved in the biosynthesis of jasmonate, and an enzyme involved in the biosynthesis of proline were strongly upregulated in both stresses. Proline is used by plants in regulating osmosis in roots. Jasmonate is an antagonist to ABA in modifying the induction of salt-stress inducible proteins (Moons et al. 1997).

Contrary to the two previous studies, another study found few genes overexpressed during salinity stress (Walia et al. 2006). Most of the identified genes were involved in the biosynthesis of jasmonate, and some genes were involved in the heat stress response (*HSPs*). This further underlines the importance of jasmonic acid as a potential "masterswitch" in salinity response in barley.

Halophytes and Extremophiles

Halophytes are plants which grow well in areas with high salt are a potential source of genes or alleles linked to salinity tolerance. There have been efforts to characterize halophytes as potentially novel of crop plants, and some plants show potential (for a review, see Glenn et al. (1999) or Shabala (2013)). One example is *Salicornia bigeloveii* (pickleweed). The plant can grow in desert-like conditions under high salt conditions and can even be irrigated with seawater (Grattan et al. 2008). Seeds of *S. bigelovii* are high in oil content, so the plant may be a potential source of cooking oil or biofuel. In content, the seed oil is very similar to safflower oil (Anwar and Bhanger 2002). Currently the genetic basis for *Salicornia*'s extreme salt tolerance is currently unknown.

In the evolution of salinity tolerance, salt tolerant grasses are a particularly interesting case. In grasses, salt tolerance has evolved a large number of times: in examining 200 halophytes at least 76 origins of salt tolerance could be identified

(Bennett et al. 2013). This indicates that grasses carry some genetic pre-requisites for the successful development of salinity tolerance. One of the pre-requisites for successful salinity tolerance is C_4 photosynthesis (Bromham and Bennett 2014), and significantly more halophytes appear in the Poaceae, and in the 'PACMAD' clade which contains all known C_4 species.

In grasses, sea-grasses like *Zostera* or *Posidonia* are of particular interest. Seagrasses have terrestrial ancestors, but during their evolutionary history seagrasses have moved 'back' to a lifestyle fully submerged in salt-water (Wissler et al. 2011). Conditions underwater with high salt concentrations and varying levels of available light resulted in strong evolutionary pressures. Positively selected genes have been identified in *Posidonia oceanica* and *Zostera marina* compared to terrestrial relatives such as *O. sativa* (Wissler et al. 2011). The GO terms "plastoglobule", "photosynthesis", "photosynthesis light reaction" and "photosystem" are the most abundant terms under positive selection in the two seagrasses. All of these GO terms are involved in photosynthesis, so the authors speculate that the lower light conditions under water have led to genomic changes in order to compensate. There is a link between the evolution of C_4 photosynthesis and salinity tolerance, which may be one factor in the successful evolution of seagrasses (Bromham and Bennett 2014). Up until a few years ago seagrasses were thought to be C_3 grasses, but one study has shown that they are actually C_3/C_4 intermediates (Andrews and Abel 1976) and there is a possibility that this contributes to their salinity tolerance.

Some seagrasses are tolerant to both a sudden increase and a slow increase of salinity (Koch et al. 2007). *Ruppa maritima* survives with an S_A of 55,*Thalassia testudinum* can survive up to a S_A of 60, and *Halodule wrightii* can survive a S_A of 65 (seawater has a S_A of 35 (Pawlowicz 2010)). However, the maximum salinity tolerance was lower when the increase of salinity was rapid, indicating that these plants need some time to build up tolerance and that extreme salinity tolerance in seagrasses is not innate. *T. testudinum*, similar to *Zostera capensis*, accumulates soluble sugars and proline and both are used for osmoregulation. This is supported by a greater surface area of the plasmalemma and a higher density of chloroplasts and mitochondria at salt transport sites. The necessary build-up of chloroplasts and mitochondria takes time, which may explain the stress observed when salinity levels rise suddenly.

Posidonica oceanica seems to be less tolerant to increased salinity concentrations than other seagrasses (Sánchez-Lizaso et al. 2008). In laboratory conditions using a P_A of 45, roughly half of exposed plants did not survive past 15 days. The genetic basis for this difference is unknown. Another study compared *P. oceanica* and *Cymodocea nodosa* in shallow and deep water and under different salinity conditions: control conditions at an S_P of 37 and hypersalinity at an S_P of 43 (Sandoval-Gil et al. 2014). In this study, *P. oceanica* exhibited lower salinity tolerance than *C. nodosa* and this may mean that different seagrass lineages have developed salinity stress tolerances independently from each other.

It would be interesting to know the underlying physiological and genomic basis of the increased salinity tolerance in seagrasses, and ultimately whether the genes responsible for can be transferred to terrestrial crop plants. However, not much is known about seagrass genomics. There are no published genetic maps or genome. Therefore, the use of seagrasses as a source of genetic tolerance to salinity for crop species is still a distant possibility.

Discussion and Future Perspectives

Rising global temperatures due to climate change and increased variation in rainfall will lead to an increase in crop stress, including heat, water and salinity stress. Many areas of salinity tolerance research are still under-developed and much work remains to be done. Future research will extend our knowledge of the genomic basis of salinity and extend into less obvious sources of salinity tolerance. Furthermore, knowledge of genetic networks involved in salinity tolerance is limited. Entire genetic networks remain to be characterized - most research takes a reductionist approach and focuses on specific genes or proteins which may not be the most productive for a highly interactive response such as salinity tolerance.

There are potential sources of salinity tolerance genes. The first is in a plant's population in the form of genetic variation. Future studies will look at even larger populations of plant accessions from more diverse backgrounds in order to find genetic variants that can be transferred to commercially available accessions in order to improve their genetic tolerance to salinity. The benefit of this approach is that it is reasonably simple and can be performed by traditional breeding. The limitation is that finding such genetic variants takes a lot of time and resources.

The second source for genetic salinity tolerance is in non-domesticated species, as summarized above for *Triticum* and *Thinopyrum*. In the future, plant breeders will continue incorporating closely related salt tolerance species in their efforts to bring salinity tolerance into domesticated species. Alternatively, breeders can identify the gene (or genetic pathways) which are involved in salinity tolerance and transfer this to the crop species using genetic transformation. The benefit of this approach is that salinity tolerant plants are easy to find. Unfortunately, research into these species has been limited the so genetic pathways and the related physiological mechanisms involved in salinity tolerance are still poorly characterized. It is thus imperative to know as much as possible about salinity tolerance pathways in diverse species so that this knowledge can be transferred to improve salinity tolerance in crops.

References

Abogadallah GM, Nada RM, Malinowski R, Quick P (2011) Overexpression of *HARDY*, an AP2/ERF gene from *Arabidopsis*, improves drought and salt tolerance by reducing transpiration and sodium uptake in transgenic *Trifolium alexandrinum L.* Planta 233:1265–1276
Andrews TJ, Abel K (1976) Photosynthetic carbon metabolism in seagrasses. Plant Physiol 650–6
Anwar F, Bhanger MI (2002) Analytical characterization of *Salicornia bigelovii* seed oil cultivated in Pakistan. J Agr Food Chem 50:4210–4214

Ashraf M, Akram NA (2009) Improving salinity tolerance of plants through conventional breeding and genetic engineering: an analytical comparison. Biotechnol Adv 27:744–752
Bennett TH, Flowers TJ, Bromham L (2013) Repeated evolution of salt-tolerance in grasses. Biol Lett 9:20130029
Bhandal IS, Mahlik CP (1988) Potassium estimation, uptake, and its role in the physiology and metabolism of flowering plants. Int Rev Cytol 110
Bromham L, Bennett TH (2014) Salt tolerance evolves more frequently in C4 grass lineages. J Evol Biol 27(3):653–659
Campos H, Cooper M, Habben JE, Edmeades GO, Schussler JR (2004) Improving drought tolerance in maize: a view from industry. Field Crop Res 90:19–34
DeRose-Wilson L, Gaut BS (2011) Mapping salinity tolerance during *Arabidopsis thaliana* germination and seedling growth. PLoS One 6, e22832
Du J, Huang Y-P, Xi J, Cao M-J, Ni W-S, Chen X et al (2008) Functional gene-mining for salt-tolerance genes with the power of *Arabidopsis*. Plant J 56:653–664
Dubcovsky J, María GS, Epstein E, Luo MC, Dvořák J (1996) Mapping of the K(+)/Na (+) discrimination locus *Kna1* in wheat. Theor Appl Genet 92:448–454
FAO (2005) Bio-physical, socio-economic and environmental impacts of salt-affected soils
Flowers TJ (2004) Improving crop salt tolerance. J Exp Bot 55:307–319
Genc Y, McDonald GK, Tester M (2007) Reassessment of tissue Na(+) concentration as a criterion for salinity tolerance in bread wheat. Plant Cell Environ 30:1486–1498
George R, McFarlane D, Nulsen B (2012) Salinity threatens the viability of agriculture and ecosystems in western Australia. Hydrogeol J 5:6–21
Giberti S, Funck D, Forlani G (2014) Δ(1)-pyrroline-5-carboxylate reductase from *Arabidopsis thaliana*: stimulation or inhibition by chloride ions and feedback regulation by proline depend on whether NADPH or NADH acts as co-substrate. New Pythol 202.3:911–919
Glenn EP, Brown JJ, Blumwald E (1999) Salt tolerance and crop potential of halophytes. Crit Rev Plant Sci 18:227–255
Gorham J, Hardy C, Wyn Jones RG, Joppa LR, Law CN (1987) Chromosomal location of a K/Na discrimination character in the D genome of wheat. Theor Appl Genet 74:584–588
Grattan SR, Benes SE, Peters DW, Diaz F (2008) Feasibility of irrigating pickleweed (*Salicornia bigelovii*. Torr) with hyper-saline drainage water. J Environ Qual 37:S149–S156
Halfter U, Ishitani M, Zhu JK (2000) The *Arabidopsis* SOS2 protein kinase physically interacts with and is activated by the calcium-binding protein SOS3. Proc Natl Acad Sci U S A 97: 3735–3740
He T, Cramer GR (1992) Growth and mineral nutrition of six rapid-cycling *Brassica* species in response to seawater salinity. Plant and Soil 139:285–294
Ilami G, Nespoulous C, Huet JC (1997) Characterization of BnD22, a drought-induced protein expressed in *Brassica napus* leaves. Phytochemistry 45:1–8
Inan G, Zhang Q, Li P, Wang Z (2004) Salt cress. A halophyte and cryophyte *Arabidopsis* relative model system and its applicability to molecular genetic analyses of growth and development of extremophiles. Plant Physiol 135:1718–1737
Ishitani M, Liu J, Halfter U, Kim CS, Shi W, Zhu JK (2000) SOS3 function in plant salt tolerance requires N-myristoylation and calcium binding. Plant Cell 12:1667–1678
Jaarsma R, de Vries RSM, de Boer AH (2013) Effect of salt stress on growth, Na+accumulation and proline metabolism in potato (*Solanum tuberosum*) cultivars. PLoS One 8, e60183
James R, Davenport RJ, Munns R (2006) Physiological characterization of two genes for Na+exclusion in durum wheat, Nax1 and Nax2. Plant Physiol 142:1537–1547
Karaba A, Dixit S, Greco R, Aharoni A, Trijatmiko KR, Marsch-Martinez N et al (2007) Improvement of water use efficiency in rice by expression of *HARDY*, an *Arabidopsis* drought and salt tolerance gene. Proc Natl Acad Sci U S A 104:15270–15275
Koch MS, Schopmeyer S, Kyhn-Hansen C, Madden CJ, Peters JS (2007) Tropical seagrass species tolerance to hypersalinity stress. Aquat Bot 86:14–24
Li B, Li N, Duan X, Wei A, Yang A, Zhang J (2010) Generation of marker-free transgenic maize with improved salt tolerance using the FLP/FRT recombination system. J Biotechnol 145:206–213

Liu J, Ishitani M, Halfter U, Kim CS, Zhu JK (2000) The *Arabidopsis thaliana SOS2* gene encodes a protein kinase that is required for salt tolerance. Proc Natl Acad Sci U S A 97:3730–3734
Ma H-S, Liang D, Shuai P, Xia X-L, Yin W-L (2010) The salt- and drought-inducible poplar GRAS protein SCL7 confers salt and drought tolerance in *Arabidopsis thaliana*. J Exp Bot 61:4011–4019
Meehl GA, Stocker TF, Collins WD, Friedlingstein P, Gaye AT, Gregory JM, Kitoh A, Knutti R, Murphy JM, Noda A, Raper SC (2007). Global climate projections. Climate change. 2007;283
Min H, Zheng J, Wang J (2014) Maize *ZmRAV1* contributes to salt and osmotic stress tolerance in transgenic Arabidopsis. J Plant Biol 57:28–42
Moons A, Prinsen E, Bauw G, Van Montagu M (1997) Antagonistic effects of abscisic acid and jasmonates on salt stress-inducible transcripts in rice roots. Plant Cell 9:2243–2259
Munns R (2002) Comparative physiology of salt and water stress. Plant Cell Environ 25:239–250
Munns R, Tester M (2008) Mechanisms of salinity tolerance. Annu Rev Plant Biol 59:651–681
Munns R, Schachtman DP, Condon AG (1995) The significance of a two-phase growth response to salinity in wheat and barley. Aust J Plant Physiol 22(4):561–569
Munns R, Hare RA, James RA, Rebetzke GJ (1999) Genetic variation for improving the salt tolerance of durum wheat. Aust J Agr Res 51(1):69–74
Munns R, Rebetzke GJ, Husain S (2003) Genetic control of sodium exclusion in durum wheat. Aust J Agr Res 54(7):627–635
Ohta M, Hayashi Y, Nakashima A (2002) Introduction of a Na+/H+ antiporter gene from *Atriplex gmelini* confers salt tolerance to rice. FEBS Lett 532:2–5
Oztur ZN, Talamé V, Deyholos M, Michalowski CB, Galbraith DW, Gozukirmizi N et al (2002) Monitoring large-scale changes in transcript abundance in drought- and salt-stressed barley. Plant Mol Biol 48:551–573
Pawlowicz R (2010) What every oceanographer needs to know about TEOS-10. TEOS-10 Primer 10:1–10
Quesada V, Ponce MRM, Micol JJL (2000) Genetic analysis of Salt-tolerant mutants in *Arabidopsis thaliana*. Genetics 154:421–436
Rabbani MA, Maruyama K, Abe H (2003) Monitoring expression profiles of rice genes under cold, drought, and high-salinity stresses and abscisic acid application using cDNA microarray and RNA gel-blot. Plant Physiol 133:1755–1767
Rajagopal D, Agarwal P, Tyagi W, Singla-Pareek SL, Reddy MK, Sopory SK (2006) *Pennisetum glaucum* Na+/H+ antiporter confers high level of salinity tolerance in transgenic *Brassica juncea*. Mol Breed 19:137–151
Ren Z-H, Gao J-P, Li L-G, Cai X-L, Huang W, Chao D-Y et al (2005) A rice quantitative trait locus for salt tolerance encodes a sodium transporter. Nat Genet 37:1141–1146
Reviron MP, Vartanian N, Sallantin M (1992) Characterization of a novel protein induced by progressive or rapid drought and salinity in *Brassica napus* leaves. Plant Physiol 100:1486–1493
Roy SJ, Tucker EJ, Tester M (2011) Genetic analysis of abiotic stress tolerance in crops. Curr Opin Plant Biol 14:232–239
Ruiz JM, Blumwald E (2002) Salinity-induced glutathione synthesis in *Brassica napus*. Planta 214:965–969
Sánchez-Lizaso JL, Romero J, Ruiz J (2008) Salinity tolerance of the Mediterranean seagrass *Posidonia oceanica*: recommendations to minimize the impact of brine discharges from desalination plants. Desalination 221:602–607
Sandoval-Gil JM, Ruiz JM, Marín-Guirao L, Bernardeau-Esteller J, Sánchez-Lizaso JL (2014) Ecophysiological plasticity of shallow and deep populations of the Mediterranean seagrasses *Posidonia oceanica* and *Cymodocea nodosa* in response to hypersaline stress. Mar Environ Res 95:39–61
Seki M, Narusaka M, Ishida J, Nanjo T, Fujita M, Oono Y et al (2002) Monitoring the expression profiles of 7000 *Arabidopsis* genes under drought, cold and high-salinity stresses using a full-length cDNA microarray. Plant J 31:279–292
Shabala S (2013) Learning from halophytes: physiological basis and strategies to improve abiotic stress tolerance in crops. Ann Bot 112:1209–1221

Shan L, Li C, Chen F, Zhao S, Xia G (2008) A Bowman-Birk type protease inhibitor is involved in the tolerance to salt stress in wheat. Plant Cell Environ 31:1128–1137
Shi H, Ishitani M, Kim C, Zhu JK (2000) The *Arabidopsis thaliana* salt tolerance gene *SOS1* encodes a putative Na+/H+ antiporter. Proc Natl Acad Sci U S A 97:6896–6901
Shi H, Quintero FJ, Pardo JM, Zhu JK (2002) The putative plasma membrane Na+/H+ antiporter SOS1 controls long-distance Na+transport in plants. Plant Cell 14:465–477
Slewinski TL, Anderson A, Zhang C, Turgeon R (2012) Scarecrow plays a role in establishing Kranz anatomy in maize leaves. Plant Cell Physiol 53:2030–2037
Suiyun C, Guangmin X, Taiyong Q, Fengnin X, Yan J, Huimin C (2004) Introgression of salt-tolerance from somatic hybrids between common wheat and *Thinopyrum ponticum*. Plant Sci 167:773–779
Sunarpi, Horie T, Motoda J, Kubo M, Yang H, Yoda K et al (2005) Enhanced salt tolerance mediated by AtHKT1 transporter-induced Na unloading from xylem vessels to xylem parenchyma cells. Plant J 44:928–938
Tambussi E, Bort J, Araus JL (2007) Water use efficiency in C 3 cereals under Mediterranean conditions: a review of physiological aspects. Ann Appl Biol 150:307–321
Thomson MJ, Ocampo M, Egdane J, Rahman MA, Sajise AG, Adorada DL et al (2010) Characterizing the saltol quantitative trait locus for salinity tolerance in rice. Rice 3:148–160
Tunçtürk M, Tunçtürk R, Yildirim B, Çiftçi V (2013) Effect of salinity stress on plant fresh weight and nutrient composition of some Canola (*Brassica napus* L.) cultivars. Afr J Biotechnol 10:1827–1832
Ueda A, Shi W, Nakamura T, Takabe T (2002) Analysis of salt-inducible genes in barley roots by differential display. J Plant Res 115:119–130
Utset A, Borroto M (2001) A modeling-GIS approach for assessing irrigation effects on soil salinisation under global warming conditions. Agr Water Manag 50:53–63
von Caemmerer S, Quick WP, Furbank RT (2012) The development of C_4 rice: current progress and future challenges. Science (New York, NY) 336:1671–1672
Walia H, Wilson C, Wahid A, Condamine P, Cui X, Close TJ (2006) Expression analysis of barley (*Hordeum vulgare* L.) during salinity stress. Funct Integr Genomics 6:143–156
Wang W, Vinocur B, Altman A (2003) Plant responses to drought, salinity and extreme temperatures: towards genetic engineering for stress tolerance. Planta 218:1–14
Wang M-C, Peng Z-Y, Li C-L, Li F, Liu C, Xia G-M (2008) Proteomic analysis on a high salt tolerance introgression strain of *Triticum aestivum/Thinopyrum ponticum*. Proteomics 8:1470–1489
Warrick M (2006) Impacts and costs of dryland salinity. Queensl Facts
Wissler L, Codoñer FM, Gu J, Reusch TBH, Olsen JL, Procaccini G et al (2011) Back to the sea twice: identifying candidate plant genes for molecular evolution to marine life. BMC Evol Biol 11:8
Wu H-J, Zhang Z, Wang J-Y, Oh D-H, Dassanayake M, Liu B et al (2012) Insights into salt tolerance from the genome of *Thellungiella salsuginea*. Proc Natl Acad Sci U S A 109: 12219–12224
Yang Q, Chen Z-Z, Zhou X-F, Yin H-B, Li X, Xin X-F et al (2009) Overexpression of *SOS (salt overly sensitive)* genes increases salt tolerance in transgenic *Arabidopsis*. Mol Plant 2:22–31
Yin XY, Yang AF, Zhang KW, Zhang JR (2004) Production and analysis of transgenic maize with improved salt tolerance by the introduction of *AtNHX1* gene. Acta Bot Sin 46:854–861
Yokoi S, Quintero FJ, Cubero B, Ruiz MT, Bressan R, Hasegawa PM et al (2002) Differential expression and function of *Arabidopsis thaliana* NHX Na+/H+ antiporters in the salt stress response. Plant J 30:529–539
Zhang HX, Blumwald E (2001) Transgenic salt-tolerant tomato plants accumulate salt in foliage but not in fruit. Nat Biotechnol 19:765–768
Zhang HX, Hodson JN, Williams JP, Blumwald E (2001) Engineering salt-tolerant *Brassica* plants: characterization of yield and seed oil quality in transgenic plants with increased vacuolar sodium accumulation. Proc Natl Acad Sci U S A 98:12832–12836
Zhu J-K (2001) Plant salt tolerance. Trends Plant Sci 6:66–71
Zhu JK, Liu J, Xiong L (1998) Genetic analysis of salt tolerance in *Arabidopsis*. Evidence for a critical role of potassium nutrition. Plant Cell 10:1181–1191

Index

D. Edwards, J. Batley (eds.), *Plant Genomics and Climate Change*,
DOI 10.1007/978-1-4939-3536-9

Printed in the United States
By Bookmasters